For Pat and her family

# Contents

3–

# LEAD PENCIL MINER

*a search for yukon gold*

Jack C. Fisher

Jack Fisher

Alamar Books · La Jolla

Author's Note:

**Multiply 1898 dollars by twenty for 1998 equivalent**

# THE SEATTLE POST-INTELLIGENCER.

VOL. XXXII. NO. 62. SEATTLE, WASHINGTON, SATURDAY, JULY 17, 1897. EIGHT-PAGE EDITION.

# LATEST NEWS FROM THE KLONDIKE.

## 9 O'CLOCK EDITION.

## GOLD! GOLD! GOLD! GOLD!

### Sixty-Eight Rich Men on the Steamer Portland.

### STACKS OF YELLOW METAL!

### Some Have $5,000, Many Have More, and a Few Bring Out $100,000 Each.

### THE STEAMER CARRIES $700,000.

### Special Tug Chartered by the Post-Intelligencer to Get the News.

The Latest Reports From the New Eldorado Arrive This Morning—Interviews With Those Who Have Come Down From the North With New-Found Fortunes—The Recent Strikes Seem to Be as Rich as Reported—There Is Plenty of Gold, But Only the Hardy and Provident Can Secure It—No Man Who Is Without a Suitable Outfit Should Tempt Fortune in That Remote Region—There Will No Doubt Be a Great Rush for the New Discoveries, and the Majority Will Outfit in and Leave From Seattle.

BRINGING BACK GOLD.

THE LAND OF GOLD.

# PROLOGUE

## *Discovery in Whitehorse*

One hundred years have passed since my wife's grandfather, Robert Harris, set out for the Yukon to find his fortune. Pat can recall vividly the many stories told by her mother, her aunts and uncles, and especially her grandmother, Keturah, who outlived her husband by thirty years.

Robert Harris and Keturah Whitney had grown up on the northeast shore of Lake Simcoe in Atherley, Ontario. Because of the economic depression that followed the "Panic of 1893", Robert and his brothers, Arthur and Thomas, sought opportunities in the west. First they helped build the Canadian Pacific Railroad, then worked their way west to salmon fish in British Columbia. Keturah eventually joined Robert in Vancouver.

On July 17, 1897, news of an important discovery reached Seattle. "A Ton of Gold!" shouted the special edition headline of *The Seattle Post-Intelligencer*, even before the steamer *Portland* docked. Tipped-off hours before, the newspaper's editor had dispatched his reporters by tugboat to get a story from the 68 miners aboard. Now, as the crew tossed lines to the dock, an excited crowd caught sight of wooden crates, metal tins, and cloth sacks on deck. Each was filled with gold: nuggets, flakes, or dust panned from tributaries of the Klondike River. Within hours of the *Portland's* arrival, word of the strike spread throughout the continent by telegraph. Within weeks, discovery of gold in the Yukon Territory was known throughout the world.

This first shipment alone was initially valued in excess of one million dollars, more than $20 million in 1998 dollars, most of it injected suddenly into a slowly recovering economy of a struggling nation. The impact of future shipments of

Yukon gold would be felt worldwide. News of previous gold discoveries had never been disseminated by telegraph. Expanding American and Canadian railroad networks would soon carry a stampede of gold seekers into San Francisco, Seattle, and Vancouver.

Thousands of men and women left their jobs and families to set out for the Yukon. Most were completely unprepared or ill-equipped for the arduous terrain and arctic conditions awaiting them. Robert Harris and his brother, Arthur, on the other hand were far better suited for the challenge. Both had spent their youth on the lakes of Ontario. They were entirely familiar with small boats: how to build them, how to pilot them. Furthermore, having endured Canada's frigid winters, the Harris brothers knew how to survive the severest of climatic conditions.

In 1987, Pat and I embarked on our first Alaskan holiday. One of our goals was to find some trace of her grandfather's Yukon adventure. First we stopped in Skagway, the principal entry point to the Yukon region. Here and at nearby Dyea, the stampeders had disembarked ninety years before, then proceeded over the Chilkoot or White Pass Trails to the Yukon River which carried them northward to Dawson City and the gold-laden Klondike tributaries nearby.

In our rental car, we climbed out of Skagway on a modern highway within view of the railroad constructed soon after the gold rush began. Our destination was Whitehorse, capital of Canada's Yukon Territory. Once over the pass, we crossed the border into Canada. Just ahead was Lake Bennett where many thousands of men had built their boats during the historic winter of 1897-98, waiting for the ice to break up and pass out of the lakes and rivers. Fortunately, Robert and Arthur arrived before that winter freeze, earning additional capital assisting miners through the difficult rapids of Miles Canyon in the upper Yukon River. As difficult as those rapids were then, we found them tamed by a modern dam and hydroelectric plant.

Entering Whitehorse, we passed a building that housed the Yukon archives. After registering at a nearby hotel, we returned and approached a research archivist. Within moments, she placed on a table before us one of the original registers of mining claims from 1897. Then to our amazement and delight, she found several claims listed for R.A. and A.J. Harris. The legend was true: we'd found the trail of our family heroes.

What we wanted to do at that point was stand and shout "Eureka!" but managed instead to maintain proper library

*Examining 1897 claims for Bonanza Creek in Yukon Archives, Whitehorse, Yukon Territory.*

decorum. We celebrated our discovery that night and returned the next day to examine the many 'creek books' for other claims made in the Harris brothers' names. Within a few hours, we identified a pattern of claims, purchases, and subsequent sales, including two on the famed Bonanza Creek. My wife recalled a line from a poem written by Robert to Keturah during that long winter:

*"I'm a lead pencil miner, no pick and shovel for me."*

Indeed, the fortune that Robert brought back to Ontario seemed to be based largely on selling rapidly appreciating property rather than the backbreaking task of harvesting gold from the frozen soil. This at least was our initial hypothesis.

Our discovery that day in Whitehorse stimulated a further search for Robert and Keturah's story. Later journeys would bring us back to Vancouver, Whitehorse, Dawson City, and the nearby gold fields. Subsequent family reunions in Atherley, Ontario revealed more of the Harris family adventure which we thought deserved to be uncovered just like the gold concealed in the Yukon. Our only obstacle was the passing of nearly one hundred years. Robert and Arthur had not allowed any barriers to halt their progress. Neither would we!

# 1

# MAN'S ETERNAL QUEST FOR GOLD

***"Put forth thy hand. Reach at the glorious gold!"***
*William Shakespeare, Henry VI*

What is it that drives men to search for gold? For many thousands of years, simple barter was the accepted means for exchange, whether for services, material goods, or food. By the fourth millennium BC, the exchange of scarce metals, primarily copper, silver, and gold was becoming more common. Ancient Egyptian kings learned how to fashion gold into molded bars, either for storage or transfer.

When mankind began to rise above a basic level of subsistence farming and the population of towns and cities increased, a more convenient medium of exchange was needed. Metallic coins seemed the best choice. Their use started first in China around 1000 BC. Four centuries later, Lydia, a small but wealthy state in Asia Minor, introduced for commercial exchange small lumps of electrum, a mineral resource abundant along its river shores.[1] Electrum is a natural alloy of gold and silver. Primitive coins were initially molded by hand but later stamped with identifying seals. This process was soon copied by the Greeks.

However, problems soon developed. Electrum could vary widely in ratio of silver to gold. People quickly recognized that some coins were richer in gold, others in silver, but the value of gold was always considered greater than silver. Lydia's final ruler, Croesus, solved that problem. His government was the first to certify and regulate the content of its coinage.[2] Furthermore, Croesus established the world's first bimetallic monetary system with one pure gold piece being equal to the value of twenty silver pieces.[3]

Lydia fell to the Persian Empire soon after, but the exchange of coins persisted and eventually spread throughout the world. Silver coins were used most commonly over the next several centuries. The Bible reports that Judas betrayed Jesus for thirty pieces of silver.[4]

Rulers learned it was necessary to restrict the distribution of valuable silver coins among ordinary citizens because they pared the edges to accumulate a personal treasure of shavings. Eventually molded ridges at the edge of coins served to discourage their devaluation.[5]

People of all social classes eventually realized that gold itself, found pure in nature, soft, malleable, resistant to corrosion, and entirely pleasing to the eye, was the preferred symbol of wealth. Rulers and governments placed so much importance on gold that ordinary citizens did too. An Egyptian artist depicted the panning of placer gold on a tomb in 2900 BC. Placer gold is a deposit of mineral that is liberated from rock by the action of water. Because gold that is found in nature is nineteen times denser than water, seven times denser than rock, it settles to the bottom. Placer gold discoveries would in time set off most of the major gold rushes in history.

During medieval times, pure gold became the common standard of exchange between nations as the basis for all international exchange. Gold was used to pay for the crusades and many subsequent wars. Gold helped Europe evolve from the dark ages of walled cities into nation states anxious for trade. Merchants began giving contractual pledges in terms of guaranteed payment in gold.

By the beginning of the fifteenth century, European demand for gold far exceeded available supply. Throughout Europe, mines long abandoned by the Romans were being reworked, but with only modest yield. This led Europe's rulers to hire alchemists for the purpose of creating gold from less valuable metals. During the fifteenth century, Charles IV of Bohemia kept a stable of alchemists working long hours to attempt conversion of lead into gold.[6]

With the advent of larger sailing vessels and new methods for navigating beyond sight of land, the freedom to seek treasure along the western shore of the African continent was taken by a few stalwart adventurers willing to risk their lives for gold. Portugal at that time was the dominant seafaring nation in Europe. Lisbon, its largest port, faced the great uncharted Ocean Sea and served as the center of exploration.[7] Portuguese prince, Henry the Navigator, embarked on a series of mid-fourteenth century voyages during which he captured the Canary Islands off northwest Africa, and many other coastal ports.

Henry commissioned his captains to depart Lisbon each spring in vessels loaded with horses, cloth, glass beads, and other products of European craftsmen. They returned the following fall with their holds filled with gold, ivory, and spices, along with slaves. The event that forced westward exploration more than any other was the fall of Christian-held Constantinople to the followers of Mohammed in 1453. With the east and its treasures no longer accessible to Europeans, the demand for new shipping routes became critical.

Enter Cristoforo Colombo. This Genoese seaman first sailed to the northern African coast in 1483 and returned with small quantities of gold. Familiar with the innovations of Henry the Navigator, Columbus had also read Marco Polo's description of the orient's "roofs of gold" and he remembered Ptolemy's prediction that gold would be most easily found in the more torrid and dangerous lands. Columbus was eager to sail due west from Portugal in the search for gold. But, he couldn't go anywhere without financing.

Unable to find the necessary capital in his native Genoa, in England, or even in Portugal, Columbus sought backing from the King and Queen of Spain. Eventually he won favor because of their acute need for gold and a matching desire to compete effectively with Portugal. The Spanish rulers offered him a ten percent commission on his yield, plus the title "Admiral of the Ocean Sea". To obtain seamen, Columbus promised them houses with tiles of gold. The search was on.[8]

We know that Columbus discovered a completely different land in a world that he and others mistakenly believed to be smaller than it was.[9] And, regrettably, his material success was limited despite three very long voyages. On the first, he encountered Caribbean Island Indians wearing gold jewelry, but no source was found. On the second, enough gold nuggets were found to lead one joyous sailor to record in his diary: "All of us made merry, not caring any longer about spicery, but only for this blessed gold". After the third and final voyage, the harbormaster in his homeport ridiculed Columbus and his party when he learned how little gold had actually been found.[10] Nonetheless, Spain's rulers were sufficiently enticed that they later backed Hernan Cortes in a further quest for gold in a land that became Mexico.[11]

Centuries later with the discovery and exploration of North America, men fell victim to dreams of gold that had not yet been discovered. Numerous early attempts to mine gold in America yielded only modest reward. Jamestown Colony settlers mistook common pyrite (fool's gold) for the genuine article, leading Benjamin Franklin to declare in 1790 that, "Gold and silver are not the produce of North America which has no mines". He was proven wrong nine years later when small discoveries were made in North Carolina. In 1829, America's first gold rush took place in Georgia when 20,000 gold seekers arrived, "all acting like crazy men", according to one observer. But there wasn't any gold! A French bank sold shares in the Mississippi Company, based on deposits of gold presumed to lie within the Louisiana Territory of North America. Thousands lost their entire investment without recourse.[12]

Americans have suffered many financially unsettling moments during their history: first in 1792, then five times in the nineteenth century, sometimes referred to as the "panics" of 1819, 1837, 1857, 1873, and 1893.[13] During these years, our nation experienced two major gold discoveries: the California gold rush in 1848-49, and a rush for gold in Colorado a few years later. What happened in California and Colorado established precedent for people to abandon their homes and take a chance for the promise of gold. Although the Yukon gold rush in 1897-99 took place in Canadian territory, many more Americans participated than Canadians.

Thoughts of discovering gold had probably never crossed the mind of John Sutter. He sought fertile soil for crops and found what he was looking for at the junction of the American and Sacramento Rivers. There he settled, built his fort, and planted crops. About him, he gathered a number of like-minded pioneers from America and foreign lands.

One of these men, James Marshall, was sent by John Sutter to establish a sawmill in the foothills east of his fort. While deepening a millrace on January 24, 1848, Marshall spotted a flash of brilliance at the bottom of the water-filled ditch. The object retrieved was in Marshall's words, "one half the size and shape of a pea". He knew instantly what it was. "I bet I found a gold mine!", he shouted to his co-workers.[14]

John Sutter quickly agreed with the findings and asked that nothing be said of any gold until after the sawmill was completed. But another gold-bearing stream was soon found, then another and another. Before long, newspapers in San Francisco were proclaiming the discovery. By June, 2,000 miners from that city were working claims in the gold fields, leaving merchants without business, newspapers without readers. But tiny San Francisco was hardly in jeopardy of becoming a ghost town. Its population soon swelled with new arrivals from the east and abroad.

*Ships abandoned in San Francisco harbor at the time of the California gold rush.*

On December 5, 1848, President James Polk addressed Congress, confirming the largest gold discovery in the nation's history and setting off the longest procession of wagon trains ever witnessed. The '49ers' left frontier outposts like St. Joseph and Independence, Missouri for the long trek west.

Others traveled as far as the trains could take them, then struck out overland from there. More came by sea, around Cape Horn, arriving in San Francisco's harbor, already dense with ships abandoned by crews on their way to the gold fields.

Two hundred thousand newly provisioned argonauts, as they came to be known, converged on a ribbon of land 10-20 miles wide, extending from the Tuolumne River in the south to the Trinity River in the north. Claims yielded an average of one to two ounces of gold a day at $16 per ounce. There were some reports of $250 days. Laborers drew $12 a day when city wages everywhere else were $1 a day. From 1848-1855, the harvest of gold from the Sierra's foothills amounted to $300 million!

Just eight years later, there came another strike. A few men prospecting on the South Platte River in the Colorado Rockies had collected $800 in gold over the summer of 1857.[15] When a Kansas newspaper mistakenly claimed yields of $800 per day, it set off another rush. During 1859, three larger strikes drew another 100,000 prospectors to Colorado, but few found gold. The Colorado gold rush made little impact on the nation's economy.

Silver, though a metal of great intrinsic value, never elicited the same excitement or appeal as gold. Prospectors near Virginia City, Nevada were at first annoyed by the bluish streaks in the rocky strata they believed contained gold. Eventually they realized they had discovered the Comstock Lode, the largest silver strike in United States history.

As America entered the 1890's, it was the railroads that dominated the industrial economy, but this led to rampant speculation and monumental debt.[16] Grover Cleveland, elected President in 1884 as a pro-business conservative democrat, enacted the needed fiscal restraint.[17] This produced a budget surplus but only for a short time. When high tariffs diminished U.S. trade with other nations, the stage was set for several years of economic struggle that brought the nation to its knees. Not surprisingly, Cleveland lost the 1888 presidential election to Benjamin Harrison but regained the office in 1892 just in time for the worst financial panic of the nineteenth century.

Public concern over heavy private debt, declining foreign trade, and rapidly shrinking gold reserves caused the nation to become divided on the proper monetary standard. Farmers and small businessmen with debt pushed for abandonment of the gold standard in favor of silver. The Sherman Silver Purchase Act of 1889 obligated the government to purchase with its remaining gold $4.5 million ounces of silver each and every month. It was the straw that broke the back of the stock market. Suddenly, a rush on the nations' banks came, followed by panic selling of shares on Wall Street, thus leading to financial chaos. Before it was over, 642 banks failed, along with many of the smaller railroads, and most of the new corporations.[18] Millions found themselves without jobs. And if that wasn't enough trouble for Americans to endure, foreign nations decided that it was time to cash in their U.S. Bonds! British vessels lined up in New York harbor waiting to load their holds with the nation's remaining gold reserves.

For three years the nation stumbled along, not yet aware that a vast new quantity of gold would soon be discovered.

# 2

# LOOKING WESTWARD

***"Turn your eyes to the great west, and there build up a home and fortune."***
*Horace Greeley, 1855*

Arthur, Thomas, and Robert Harris were like-minded young men, three among sixteen children sired by Collingwood Harris. Born within four years of each other, they grew up near the village of Atherley in the Province of Ontario, 90 miles north of Toronto. The boys might have adopted any of the numerous trades and ventures pursued by their enterprising father: boat builder, bridge contractor, hotel keeper to name a few. But all three seemed more inclined to explore opportunities elsewhere.

Their eyes were, in fact, turned westward to the still remote Province of British Columbia. Travel there was becoming easier due to the completion of the transcontinental Canadian Pacific Railway (CPR) in 1885. Although many hundreds of young Canadian men had been recruited to build the railroad alongside many thousands of immigrant Chinese, work continued long after initial operations began. Arthur and Thomas were among those hired to improve track and develop facilities for the profitable outcome that was still many years away.

Building the railroad was almost beyond the reach of Canada's human and financial resources. It took a nation of 40 million to complete America's railroad,[1] and Canada didn't have one-tenth that population. Worse, the railroad across Canada required an extra 1,000 miles of track. Furthermore, no American crew faced a barrier as forbidding as Canada's 700 miles of solid granite wasteland bordering Lake Superior's northern shore.[2]

Yet, in 1871, two years after completion of America's coast-to-coast railway, conservative prime minister John Macdonald pledged a transcontinental railroad to British Columbia if it would join the Dominion. British Columbia readily agreed. According to Alexander Mackenzie, leader of the Parliament's liberal opposition, Macdonald's offer was "an act of insane recklessness".[3] There were many who agreed. When the last rail spike was driven in 1885, cost overruns had caused the Canadian treasury to be severely over committed. Every Canadian citizen bore a heavy tax burden imposed by the project, which continued for the remainder of the nineteenth century and well into the twentieth.

In 1885, the same year Canada was celebrating completion of its railroad, the Harris family was mourning the death of its patriarch.[4] For 76 years, Collingwood Harris had manufactured boats and built bridges throughout Ontario. In Atherley, his was the first bridge constructed over the narrows, a short isthmus running between Lake Simcoe to the south and Lake Couchiching to the north. Enchanted by the terrain, he moved his family to Atherley and began acquiring property along the western shore facing the village. Here, he became the family's first hotel owner. Twice a widower, he was already the father of eight children before marrying Elizabeth Murphy. She bore him eight more children, among them four boys: Thomas, Arthur, Robert, and Shuter.[5] The four girls were Emma, Rachel, Ida, and Martha.[6]

In the year following their father's death, Thomas and Arthur could wait no longer to strike out and build their futures. Hired onto a CPR construction crew, they were assigned to the mountain division working Alberta's Rocky Mountain Range and British Columbia's Selkirk Range. It became Robert's task to look after the Harris women and property in Atherley.

Thomas and Arthur were not the only young men from their Township working in British Columbia. Also in the camp was William Whitney of Atherley. During that summer, William received a letter from his sister Julia in which she wrote of her special relationship with Thomas Harris. She ended by urging her brother not to stay away much longer and to bring both of the Harris boys back safely with him. Soon after their return, Julia Whitney married Thomas Harris.[7] Two years later, Robert Harris married Julia's sister, Keturah.[8]

For Thomas, one trip west was enough and he chose to remain with his family in Ontario for the rest of his life. Arthur, on the other hand, couldn't get enough of the west and soon left again, this time settling in Granville at the end of the CPR line. It was a small town on the Burard Inlet of Puget Sound, later renamed for the British explorer George Vancouver. Here Arthur established himself. Bitten by the same spirit of adventure, his sister, Ida, followed. Their letters eventually persuaded Robert, who had not yet traveled west, to join them, leaving Keturah behind in Atherley.

In the 1890's there were only a few major industries in the

*Robert Harris marries Keturah Whitney in Atherley, Ontario on July 3, 1888*

west. One of them was Pacific salmon fishing. Robert and Arthur recognized opportunity when they saw it. Arthur had already purchased a new 'gasboat'.[9] Too small to warrant steam power, its little engine ran on a crude petroleum distillate, newly available on the Pacific Coast. So proud of their gasboat, Arthur and Robert sent a picture to the family in Ontario with comments on the reverse side. "Access to fishing waters",

*Robert and Arthur Harris in their gasboat nearby Vancouver, British Columbia.*

they wrote, "was governed by the wishes of local Indians who are far less inclined to grant similar permission to Japanese fishermen now present in great numbers". The natives of Puget Sound had been guaranteed unrestricted fishing rights when they ceded their land to the Canadian Government.

Within weeks of arriving, Robert knew that he would be able to make a living for himself and began to look for land he could build a house on. Early in January, 1890, he purchased a lot from Vancouver Improvement Company Limited. The price was $300 and his real estate broker, James Holland, was able to obtain 100% financing from a Mr. Allen McDonell. The terms were eight percent interest with the entire amount due in one year.[10]

The canning industry was also growing rapidly. Prudent

entrepreneurs working the nearby waters developed as much interest in canning operations as in fishing itself. Robert and Arthur invested in a fractional ownership of the English Bay Canning Company,[11] located just south of the Vancouver peninsula.

Salmon yields were enormous in these waters but markets for the fish were thousands of miles away. Exporting pre-salted salmon in large barrels would no longer do; there was too

*English Bay Cannery near Vancouver.*

much spoilage at the point of delivery. A better storage method was needed. Shipping foods in airtight cans was an early nineteenth century Scottish innovation. By rolling thin sheets of steel, the Scots made it possible to create a cylindrical unit that could be vacuum-sealed after the contents were cooked. Canning had been used for oysters and other sea products since 1840 on the eastern coast of North America. The technology was transferred to the Pacific coast during the 1860's when a new stamping process permitted tin cans to be made cheaply enough for wider sale.[12] Salmon could now be processed immediately for shipment around the world. In England, a tin of Pacific salmon was at first considered a delicacy. Eventually it became a staple

For the Harris brothers, business was good, very good in

fact. They were involved in Canada's fastest growing industry and the nation's second largest export after grain. The brothers commitment to salmon provided opportunity, adventure, and prosperity for them and their workers. Their thoughts probably did not often turn to gold, even though word of small gold discoveries in Alaska and the Yukon Territory were sometimes discussed on the streets of Vancouver.

# 3

## "A TON OF GOLD"

***"Everybody is looking for gold . . . O if men would only work as hard for the kingdom of heaven.***
*Father William Judge, 1894*

Although the rush for Yukon gold started off with a bang when steamers sailed into the Seattle and San Francisco harbors in July of 1897, the early discoveries in that region were barely noticed. Michael Gates, Yukon historian and author of *Gold at Fortymile Creek* makes clear that nearly a quarter century elapsed between the arrival of the first white men in that area and the major discovery that took place in the summer of 1896.[1]

The earliest explorers to enter the Yukon River basin had been far less interested in gold than they were in trading fur pelts with the natives. Fur skins were plentiful and their value had already been established. The presence of gold was uncertain and its value had not yet been established. First among the settlers was Robert Campbell of the Hudson Bay Company, who found traces of gold between 1848-1852 while working near Fort Selkirk on the Yukon River.[2]

Christian salvation was another goal of the early settlers. The Reverend Robert MacDonald had been sent to the region in 1864 by Canada's Church of England. Although he had panned some gold on the Yukon River, missionaries and early traders could agree on one principle: they wanted the region for themselves without the intrusion of prospectors.

Arthur Harper, an Irish immigrant, held his own vision of

finding gold in the vast unexplored northwestern region. Jack McQuesten was drawn to British Columbia from New Hampshire by the lure of gold. They eventually met and settled at Fort Yukon during the summer of 1873. Harper would devote the next ten years to gold prospecting on most of the Yukon River's tributaries without much success. McQuesten, on the other hand, recognized there would be ready cash from providing prospectors with food and supplies. In 1874, he opened a trading post about six miles downriver from the future site of Dawson City, which he named Fort Reliance.[3] This became the principle commercial center in the Yukon until 1886. Other local settlements on the river were named for their distance from McQuesten's Fort Reliance, e.g. 'Fortymile' was located about that distance from the trading post.

During these years, there were only a few dozen men in the entire Yukon basin, largely because access was limited to either the Hudson Bay Company route via the Mackenzie River or else by steamer from St. Michael, an old Russian port facing the Bering Straight near the Alaskan outlet of the Yukon River. Then in the late 1870's, George W. Holt began using the Chilkoot Pass to gain entry to the region. High above the Chilkat Indian trading center of Dyea, he climbed the steep path over the pass to a chain of lakes forming the headwaters of the Yukon River. This pass was the shortest and most direct route, attracting several hundred gold seekers before 1897.

In 1882, more than seventy men used the Chilkoot Pass, among them Joe Ladue who remained that winter and many thereafter. Ladue and others prospected that year along the Deer River, known to local natives as the 'Thronduik'. A small branch of the Deer River was named Rabbit Creek. Ladue and his party camped one night on this creek, never suspecting that one day it would be renamed Bonanza Creek. Meanwhile at Fort Reliance, McQuesten recognized that the arrival of more prospectors would mean an increasing demand for goods. Accordingly, he imported fifty tons of supplies into Fort Reliance during the summer of 1885. These goods were carried by the Chilkats for $12 per hundredweight. Small river steamers also brought cargo up the Yukon from St. Michael during the summer.

As the prospecting season of 1886 drew to a close, there was a significant gold discovery on the Fortymile River. But little could be done until spring because all the rivers were frozen solid. Arthur Harper had by this time become a business partner of McQuesten's. Anticipating a surge of miners in the spring, Harper went to San Francisco, and substantially expanded an order for supplies coming to Fort Nelson, their new trading post at the junction of the Stewart and Yukon Rivers. But the unexpected 1887 rush to the Fortymile River discovery dictated the need for establishing another trading post there as well. The two partners guessed right. By midsummer 1887, steamers had carried more than 100 tons of supplies up the Yukon River to the new community of Fortymile.[4]

Meanwhile, two men were heading for the Yukon with a very different goal in mind. George Dawson, representing the Geological Survey of Canada, came from the east by way of the Mackenzie River route. William Ogilvie, a Dominion land surveyor, traveled over the Chilkoot Pass, assisted by George Carmack, his translator.[5] The arrival of government officials in the region disturbed many of the miners. By nature, they avoided authority, especially the rule of government.

Dawson and Ogilvie's findings surely did affect them. Fortymile was confirmed to be east of the 141st longitudinal meridian which defined the Canadian-Alaskan border. That meant the gold at Fortymile was clearly within Canadian territory. Since most of the miners were American, any gold recovered by them was subject to customs duties, not that the government was positioned to collect duties or to enforce any

Canadian laws. At Ogilvie's recommendation, the government did not intervene because it might drive the Americans westward into Alaska, thus limiting exploration of Canadian land.

Over the next five years, mining continued at Fortymile River and by 1893, miners had taken out gold worth $400,000. Unwilling to leave profitable claims, they began to work year round. From the beginning, placer mining, the sifting of dense gold particles from less dense rocky debris, was the preferred method because it was easiest for beginners to pursue without enormous investment of capital. At Fortymile, miners initiated a variation called 'drift mining' which began with a thawing of frozen ground with enormous wood fires. While the ground was soft, they dug vertical pits and horizontal shafts. In 1893, wood for the fires was plentiful because the surrounding hillsides were still covered with forests. All the rock and soil dug out of the ground during winter was set aside in drifts. In the spring, flowing water allowed for sifting of those drifts and the extraction of gold. During the winters, the more fortunate prospectors were able to identify visible veins of gold in the rock which they called their 'paystreak'. Drift mining was back breaking work: thawing, then digging, inch by inch, foot by foot. But there was little else to do during the freezing winter.

As summer approached, there came a risk of collapsing tunnels as melting snow and warming soil caused them to give way. Even though lives were lost in the cave-ins, drift mining was considered an important technical advance. The work season was now longer, and the identification of paystreaks meant greater yield per ton of soil. Prices for necessities rose too. Laborers demanded $10 in hourly wages, flour sold for $19 per hundredweight, a pair of boots cost $18. Back in Seattle, laborers were still working for $1 a day and buying boots for $2.

With more and more prospectors arriving, the Canadian government became concerned about law and order in the Yukon Territory. A contingent of Northwest Mounted Police was dispatched from headquarters in Regina, Saskatchewan, under the command of Charles Constantine. One of their tasks was to begin collecting custom duties on the gold being taken out by the largely American population of miners. That first year, the Mounties collected $3,249. In addition to their annual wage of $1,000, a NWMP officer was given a ten percent commission on all duties collected.[6]

When the Mounties arrived, Fortymile was already a thriving town with saloons, bakeries, restaurants, blacksmiths, a theater, two doctors, two missions (one Catholic, the other Anglican), as well as a number of very busy prostitutes. The first women to enter the Yukon had been the wives of missionaries. Other women arrived looking for missing sons or husbands. As always, a careful social distance was maintained between white women and native, good women and bad.

White men in the Yukon now exceeded 1,000, and their numbers necessitated expansion of the local NWMP contingent. Health and safety had become another part of their responsibilities. Untended lamps in buildings were forbidden because they were known to cause fires. During winter months, Mounties also enforced daily consumption of lime juice to prevent scurvy.

By 1896, the initially meager findings of gold on Birch Creek near Circle, Alaska, had grown so steadily that it replaced Fortymile as the region's gold mining center. As one community gave way to the other, buildings were dismantled and the components transported to the next location for re-use. It was a simple task to float these logs downstream from Fortymile to Circle, 200 miles closer to the Yukon River's outlet to the Bering Sea, where steamers arrived frequently from St. Michael.[7] New warehouses were built to receive supplies for the Alaska Commercial Company and other importers. While standard payment for goods and services continued to be gold, credit was often extended to maintain a miner's 'grubstake' in anticipation of a profitable gold harvest.

Who extended this credit? Individuals like Harper,

McQuesten, and newly arrived Joe Ladue, all competing for the prospector's business. Initially McQuesten had been king in Circle, but not for very long. Just as quickly as Circle reached its pinnacle, it would fade. Soon all eyes and every boat would turn 250 miles to the southeast toward the Thronduik River,[8] pronounced 'Klondike' by the prospectors.

The principle players in the big strike on Rabbit Creek were George Carmack and his two Indian partners, Skookum Jim and Tagish Charlie. The supporting cast included Robert Henderson, another miner, and Joe Ladue, who had previously grubstaked Henderson and employed Carmack. Ladue, more than any other businessman in the Yukon, would exploit the great discovery for his own benefit by founding Dawson City.[9]

Back in 1885, Carmack had come to the Yukon from California. Initially he had worked out of the Alaskan villages of Juneau and Dyea as a seaman, then guide, and eventually prospector. Spotted by surveyor William Ogilvie, Carmack was selected as a guide because of his familiarity with the language and customs of the native population. Carmack also hauled freight over the Chilkoot Pass, trapped furs and mined gold at Fortymile. When he was out of funds, he worked at Ladue's trading post near Sixtymile, a small community that would later be renamed for Ogilvie.

After Carmack married Skookum Jim's sister, whom he renamed Kate, he accepted a native lifestyle. Fellow miners referred to him as Siwash George, after the derogatory term for local Indians. Now at the onset of the summer of 1896, Carmack was deciding where he would prospect next. Flipping his last silver dollar, it came up heads meaning he would head down river to Fortymile. After filling his boat with supplies, he poled his way back upriver to the mouth of the Klondike where Kate, Skookum Jim, and Tagish Charlie were waiting at their meadow campsite. There, in early August, they encountered Robert Henderson for the first time.

*George Carmack*

*Skookum Jim*

*Tagish Charley*

*Robert Henderson*

Henderson, a Nova Scotian, was smitten early with the gold bug and it captured his interest for a lifetime. Regrettably, he lacked both the skill and the luck needed to achieve the fortune he desired. Moving from one gold discovery to the next, he usually arrived long after it peaked. Attracted to the Yukon by the activity at Fortymile, he again arrived too late. However, there were new rumors everyday and Ladue, willing to grubstake him, urged Henderson toward the Indian River. He spent the long winter of 1894-95 there, thawing and digging his soil. But when spring came and the water flowed again, he found little gold in his drifts. Worse, he had severely injured his leg and lost nearly forty pounds while nursing the wound. When physically able, Henderson restocked and came to the mouth of the Klondike where he spotted Carmack's party. Their encounter would further isolate him from success.

Henderson was heading for Gold Bottom Creek on the other side of the ridge from Rabbit Creek where Carmack planned to work. Ordinary mining courtesy held that prospectors working areas close to one another would remain in contact, sharing news of their findings. But Henderson offended Carmack at their first meeting, expressing his disdain for Carmack's Indians. A few days later, Carmack, Jim and Charley climbed a ridge, saw what they thought was the smoke of Henderson's campsite, and walked down to see how he was doing. Noticing Henderson's generous tobacco supply, Carmack offered to buy some. Henderson arrogantly refused, unwilling to "share tobacco with any Siwash". Now angry, Carmack and his friends returned to Rabbit Creek. It was more than a discourtesy; it was a serious error Henderson would regret for the rest of his life.

On August 17, 1896, a date later designated 'Discovery Day' in the Yukon, Skookum Jim paused beside Rabbit Creek for a drink. As he knelt, he spotted a glimmer of gold on its bottom and shouted for George and Charley. There, in plain view, they all saw more gold in one place than they had ever seen before. It was a paystreak, fully revealed by the passage of running water. Carmack didn't waste a moment. He took a pencil from his pack and wrote these words on the trunk of a nearby tree:

*"To whom it may concern: I do this day locate*
*and claim, by right of discovery five hundred*
*feet, running upstream, by this notice.*
*Located this 17th day of August, 1896.*
*G.W. Carmack".*

This made Carmack the registered discoverer, and entitling him to a 500 foot claim above and another below the point of discovery. For Jim, he claimed the next 500 feet above (called #1 Above Discovery), and for Charley, the next 500 feet below (#1 Below Discovery). This was the accepted procedure for designating claims according to 500 foot land segments above and below any discovery point. All claims extended 200 feet on either side of a creek, making a typical claim 2,000 square feet.

Carmack and his partners quickly headed toward Fortymile to register their claims. Along the way, they announced their discovery to every miner they saw but not to Henderson, who for the rest of the season remained oblivious to what was happening just over the ridge. Within days of the news, Fortymile lost all of its inhabitants. Every possible claim on Rabbit Creek was taken within a few days. On August 22, twenty-five claim holders met and voted to rename Rabbit Creek. It would thereafter be known as Bonanza Creek, and its smaller tributary became Eldorado Creek. Within a month and a half, 338 claims were registered on these two creeks alone.

Because he made it his business to always know what was happening in the region, Ladue folded his entire operation in Ogilvie and brought everything he owned downriver to the patch of meadow at the junction of the Klondike and Yukon

Rivers where Henderson had first offended Carmack. Anticipating the inevitability of a new town there, Ladue staked out 160 acres in his own name; 18 more acres were reserved for government use. While the miners dug and sweated and panned, he re-established his sawmill and built a new trading post. It wasn't long before miner's tents began appearing.

Carmack and his two Indian sidekicks were each taking out about five ounces per day from their claims, enough to pay off all back accounts and purchase more supplies for the coming winter. And what a winter it was! Dawson City grew steadily as the news spread westward into Alaska and as far south as Juneau.[10] That winter, 1,500 men worked the nearby creeks, first thawing the soil, then digging shafts. The gold was obvious even to bystanders who could witness sparkling in the soil as a miner worked in his hole by candlelight.

By April, 1897, another thousand men had entered the region on their way to Circle City, but the new rumors persuaded most of them to head for the Klondike River as soon as the river opened. They had all subsisted that winter on a diet of beans ("Yukon strawberries") and sourdough bread ("as dense as the gold they were mining").[11]

The ice finally broke on May 14. Huge cakes, some as large as a house, began to flow northward towards Alaska and the Bering Sea. Several days later, the ice had thinned enough for small boats to navigate safely on the Yukon. Soon afterward, newcomers appeared from around a bend above Dawson City. Awaiting them was an expanding tent city filled with men anxious to receive any news from the outside world. The Pope was well, Queen Victoria still ruled, no wars were anticipated. Most important to the miners, however, was the fact that Gentleman Jim Corbett had been knocked out by his challenger![12]

Now that water was again flowing in the Yukon region, every pan, every sluicebox was in use day and night. The richness of the claims could be fully comprehended only after separation from the soil. The infusion of newly-mined gold into Dawson City created the classic conditions for price inflation: too much money chasing too few goods. Log cabins went for $1,000 each. Bacon sold for seven times its ordinary value. Salt was so scarce that it was bought and sold for its weight in gold. And dance hall girls were getting $100 for their services.[13]

As new prospectors arrived, eighty of Dawson City's newly rich were waiting to embark. Around them were stacked their personal goods and recently acquired wealth, packed tightly into suitcases, sacks, even Royal Baking Powder tins. Not until June was the whistle of the Alaska Commercial Company's sternwheeler *Alice* heard. Then the whole town turned out to welcome its cargo of provisions, especially the fresh fruit and vegetables. Two days later, the *Portus B. Weare* arrived, its hold filled with even more supplies, although just a fraction of the tonnage required to get Ladue's operation through the hectic season ahead of him.

The time came for those leaving to hoist their property aboard the two vessels. Each passenger brought wealth ranging from $25,000 to $500,000. The ship's safe was full and its position had to be reinforced with timbers because of the extra weight. Departing Dawson City, they began the 1,700 mile journey down the Yukon River.

On June 27, both ships arrived in St. Michael, a barren place according to most observers of the day. There, two large ocean steamers awaited them: the *Portland*, bound for Seattle, and the *Excelsior*, bound for San Francisco. Because the *Portland* was scheduled to leave first, most of the miners boarded it, less concerned about their destination than getting back first. However, the *Portland* took the slower inside passage, while the *Excelsior*, keeping to the faster ocean route, arrived in San Francisco July 15, a full two days ahead of the *Portland's* arrival in Seattle. Initial response to the miner's stories was one of skepticism. But enough speculation arose that the local editor telegraphed *The Post-Intelligencer* in Seattle to expect arrival of another gold-laden vessel.

*Steamer Excelsior arriving in San Francisco Harbor July 15, 1897, two days ahead of the Portland's arrival in Seattle..*

# 4

# HEADING FOR THE YUKON

***"What a mad rush this is to a land nobody knows anything about."***
*Joseph Grinnell, Stampeder*

The stampede for Yukon gold was unlike any previous gold rush in history. No communication network or transcontinental railway had existed in 1848 when prospectors headed for Sutter's Mill. Back then, nothing in the world traveled faster than a galloping horse.[1] But by 1897, a telegraph system spanned the continent carrying messages at the speed of light. The rapidly expanding railroad network was delivering passengers and freight to every major city on the Pacific coast. So news of the steamer *Portland's* arrival in Seattle harbor with an estimated "ton of gold" re-ignited a new lust for gold. No news had ever before spread so quickly with as much impact.

The Harris brothers didn't take long to make their decision; they were ready to join the migration northward within a month of hearing the news. This meant leaving their fishing and canning enterprise behind for others to look after. And it meant leaving their families behind as well. They didn't hesitate for a minute.

Throughout the world, the characteristic response was spontaneous, immediate, and unwavering. John Nordstrom, a 26 year-old Swedish immigrant working near Seattle as a jack-of-all-trades, later achieved success and fame as a clothing retailer. But on that Sunday morning he read *The Seattle Post*

-*Intelligencer's* leading story, jumped up and exclaimed, "I am going to Alaska!" Indeed he did. Writing later about the experience, he described packing everything he owned that very afternoon. Taking what little money he had, he boarded a train for Seattle and a new adventure.[2]

In Seattle the next morning, Nordstrom learned that every berth on the *Portland's* return voyage had already been sold. Responding to the heavy demand, steamship companies began to reroute many of their vessels to the Alaskan ports. Nordstrom and two newly found partners managed to secure tickets on the steamer *Willamette*, a coal carrier that customarily ran between Seattle and San Francisco. It took a week for the conversion that would allow this ship to carry men, equipment, and horses to the ports of Skagway and Dyea. Overhearing someone ask the difference between first and second class, Nordstrom learned that, "In first you sleep with horses, in second with mules."[3] There were 600 of each on board. He chose second class and later reported sleeping with both! There were also 1,200 men and their supplies, the largest boatload yet to sail for Skagway. The *Willamette* departed Seattle on Monday, August 2, 1897.

On that same day in San Francisco, the steamer *City of Topeka* departed for Alaska. On board was young Jack London. Six years later, he was to write his most famous story, *Call of the Wild*, based on his experiences in the Yukon. London and his party landed in Juneau on August 2nd, then transferred to a smaller vessel for the remaining one day journey to Skagway. He reached Chilkoot Pass by August 12, just as Robert and Arthur Harris, back in Vancouver, were packing for their own departure.

Far away in New York City, reactions to the gold discovery were mixed. Knowledgeable financiers consulted mining engineers to determine whether the quartz content of the soil near Dawson City was sufficient to make this a significant discovery.

Without more geologic data, few investors were willing to risk their capital. Meanwhile, less informed clerks were abandoning their desks and arranging for passage westward. Sailors were jumping ship and heading for railway offices to purchase their transcontinental tickets.

Back in Vancouver, the Canadian Pacific Navigation Company announced that its steamship *Islander* would now travel the Victoria-to-Dyea route because of extraordinary demand.[4] Had Robert and Arthur wanted to leave as early as the *Islander's* first journey in late July, they would have been among the many men turned away. But they did obtain passage on that boat's next departure scheduled for August 15th. They realized more than most people that little time remained before the onset of winter and all the problems that would present.

Also booking passage for the *Islander's* mid-August departure was Tappan Adney, a veteran reporter for *Harper's Weekly*. He purchased his ticket on July 30 at the New York office of the Canadian Pacific Railway. This gave him very little time for preparations and a transcontinental crossing.

Adney's editors directed him to proceed to the Yukon and report on an event they believed would captivate a worldwide audience. His reports were to be carried by the *London Chronicle* as well. Adney was not the only correspondent dispatched to the northwest; journalists were also sent by *Scribner's Magazine*, *The Illustrated London News*, and *The Century*. It would be Adney, however, who met two remarkable Canadian brothers on the *Islander*. Observing their ingenuity, he reported on their boat building and piloting skills.[5]

Newspaper accounts of the estimated value of gold still undiscovered in the Yukon region now exceeded $35 million, a figure based more on arbitrary speculation than upon geologic survey. The 68 miners arriving on the *Portland* had submitted their gold for assay, and its valuation was determined to be $800,000, somewhat less than original estimates. Competition for space on trains and steamers was producing chaos, disrupting not only the northwest shipping routes but also those throughout the world.

As early as July 23, Vancouver and Seattle newspapers began to warn gold seekers that each was going to need more than 1,000 pounds of supplies and equipment to survive a Yukon winter. In February, 1898, the NWMP would establish regulated entry points where each traveler was required to show sufficient provisions. Without this requirement, many more would have perished in the Arctic climate.

Commercial outfitters in Seattle were already harvesting enormous profits,[6] as were their counterparts in San Francisco and Vancouver. Trains unloaded provisions nearly every day, along with many hundreds more men and a few women, most bound for the gold fields. By late autumn of 1897, more than 30,000 had boarded northbound steamers.

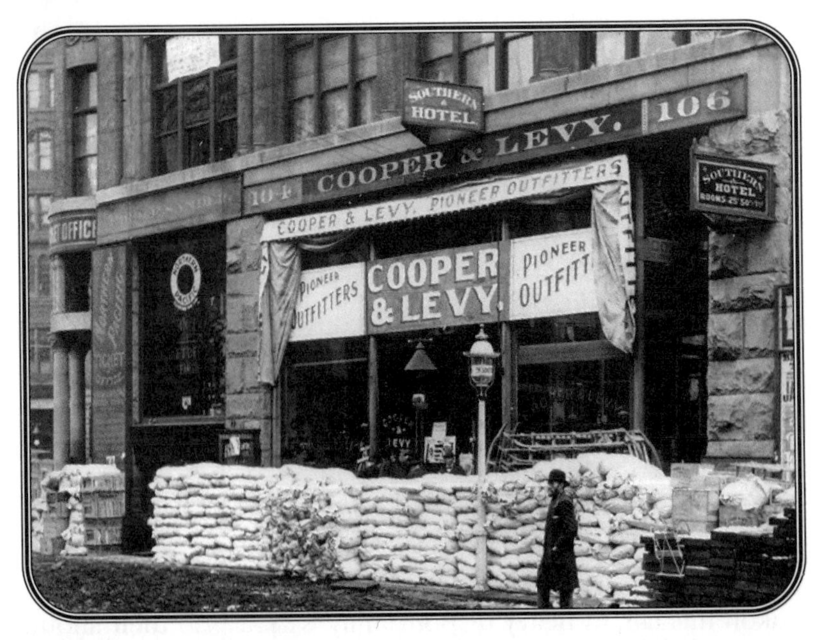


*Supplies bound for the Yukon stacked in front of a Seattle store.*

For those determined to leave civilization behind, there was a choice of routes. The most commonly taken course was by steamer up the inland passage to either Dyea or Skagway in southeastern Alaska, then by foot over one of the two available mountain passes, the Chilkoot or the White. From there, travelers continued by boat through a descending chain of lakes to the Yukon River which flowed northward to Dawson City, 2,000 miles from Seattle.

A second alternative, for those with more time and money, involved booking passage on more substantial ships capable of traveling the Alaskan coast to St. Michael on the Bering Sea, then transferring to a smaller draft vessel for passage up the Yukon River to Dawson City. But this route was only appropriate for men leaving by mid-summer because the Yukon started to freeze in September, making Dawson City an unattainable goal. Although it was sarcastically billed as the 'clean fingers' route by those without the funds, many who took it got stuck along the way and suffered as much as those who braved the overland trail. Several hundred travelers were in fact trapped that year, spending a frigid winter in tiny Circle City, Alaska. Others were marooned in St. Michael or forced to turn back.

The third, referred to as the all-Canada route, was selected by many Canadians but only a few Americans. By train, they rode to Edmonton in the Alberta Territory, then by foot northward along the Mackenzie River, continuing westward over uncharted landscape to Dawson City. This was the most demanding route of the three. A few travelers were successful; many more lost their way and froze to death.

Voices of warning could have been heard if anyone had been willing to listen. Louis Sloss of the Alaska Commercial Company knew the territory and said, "I regard it as a crime for any transportation company to encourage men to go to the Yukon this fall. A heavy responsibility will rest on their shoulders should starvation and crime prevail."[7] The Traveler's

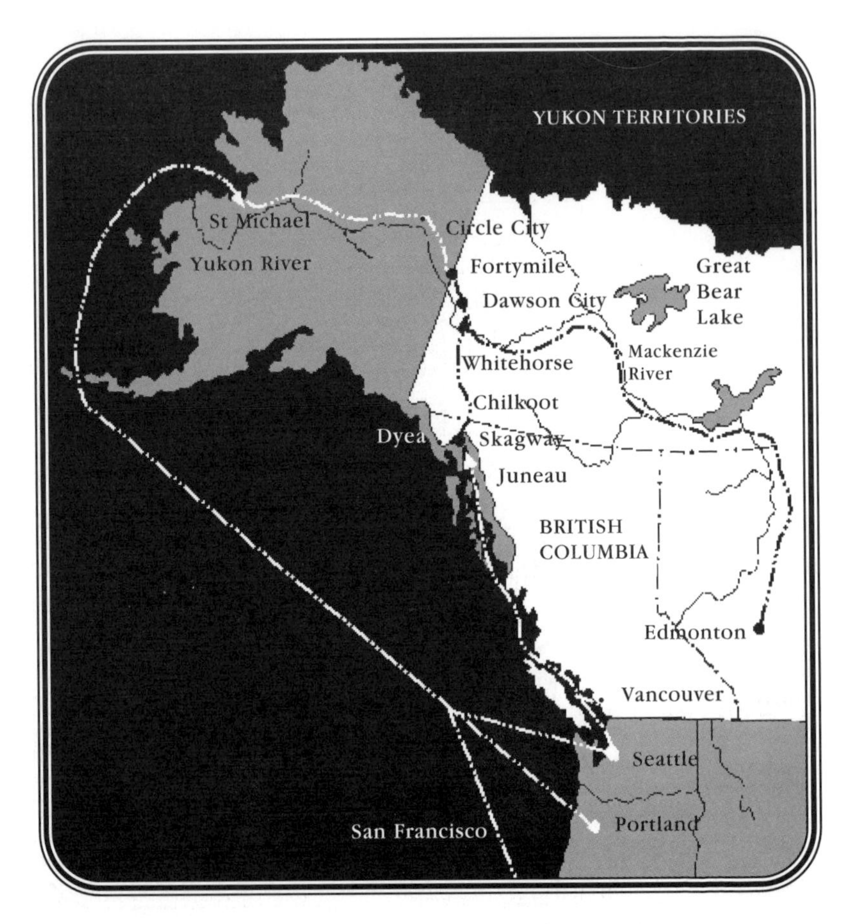


Insurance Company announced it would not sell insurance to anyone going to the Yukon. U.S. Secretary of the Interior Bliss issued a warning against any attempted travel to the Klondike that season. Canadian authorities issued similar pleas.

Nonetheless, the migration continued. Reaching Montreal, Tappan Adney telegraphed ahead to reserve additional space on the Islander for the six horses he intended to buy in Victoria. Continuing by rail to Winnipeg, he discovered that the Hudson Bay outlet was already sold out of winter clothing. The best he could find was a long, hooded coat made of black duffel cloth. Boarding the next day's train, he arrived in Vancouver on August 8th, then proceeded to Victoria by ferry.

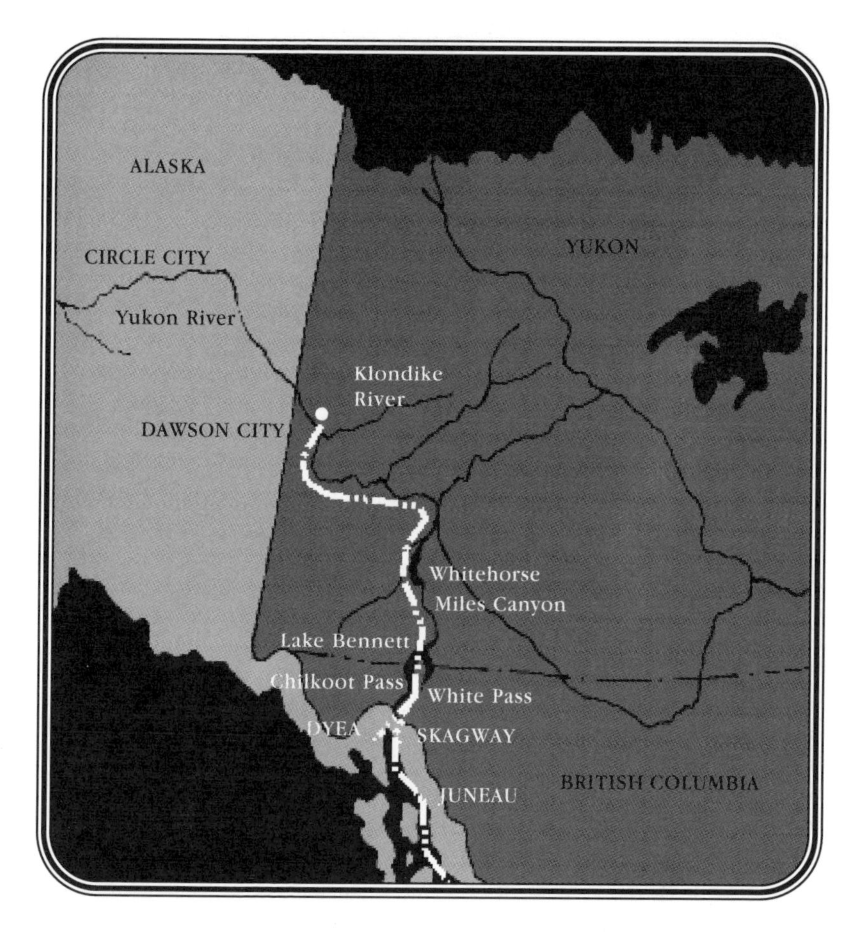


He found the streets of Victoria jammed with men intent on preparing for their journey northward. They'd come from England, Ireland, Germany, France, Australia as well as from Canada and the United States. Among them were spoken many languages, with one magic word common to all: "Klondike!"[8]

Adney found clothing sparse in Victoria, requiring ingenuity to assemble the many different articles necessary to assure him warmth and survival. Nearly every mule and horse on Vancouver Island had already been purchased for transport, whether fit for the coming hardships or not. The hay to feed these pack animals cost $240 per ton.

Adney later recorded for his readers what a man would

THE KLONDIKE STAMPEDE

8 sacks Flour (50 lbs. each).
50 lbs. Bacon.
150 lbs. Split Peas.
100 lbs. Beans.
25 lbs. Evaporated Apples.
25 lbs. Evaporated Peaches.
25 lbs. Apricots.
25 lbs. Butter.
100 lbs. Granulated Sugar.
1 1/2 doz. Condensed Milk.
15 lbs. Coffee.
10 lbs. Tea.
1 lb. Pepper
10 lbs. Salt.
8 lbs. Baking Powder.
40 lbs. Rolled Oats.
2 doz. Yeast Cakes.
1/2 doz. 4-oz. Beef Extract.
5 bars Castile Soap.
6 bars Tar Soap.
1 tin Matches.
1 gal. Vinegar.
1 box Candles.
25 tbs. Evaporated Potatoes.
25 lbs. Rice.
25 Canvas Sacks.
1 Wash-Basin.
1 Medicine-Chest.
1 Rubber Sheet.
1 set Pack-Straps.
1 Pick.
1 Handle.
1 Drift-Pick.
1 Handle.
1 Shovel.
1 Gold-Pan.
1 Axe.
1 Whip-Saw.
1 Hand-Saw.
1 Jack-Plane.
1 Brace.
4 Bits, assorted, 3/16 in. to 1 in.
1 8-in. Mill File.
1 6-in. Mill File.
1 Broad Hatchet.
1 2-qt. Galvanized Coffee-Pot.
1 Fry-Pan.
1 Package Rivets.
1 Draw-Knife. [Granite.
1 3 Covered Pails, 4, 6, and 8 qt.
1 Pie-Plate.
1 Knife and Fork.
1 Granite Cup.
1 each Tea and Table Spoon.
1 14-in. Granite Spoon.
1 Tape-Measure.
1 1 1/2-in. Chisel.
10 lbs. Oakurn.
10 lbs. Pitch.
5 lbs. 2od. Nails.
5 lbs. 1od. Nails.
6 lbs. 6d. -Nails.
200 feet 5/8-in. Rope.
1 Single Block.
1 Solder Outfit.
1 14-qt. Galvanized Pail.
1 Granite Saucepan.
3 lbs. Candlewick.
1 Compass.
1 Miner's Candlestick.
6 Towels.
1 Axe-Handle.
1 Emery-Stone.
1 Sheet-Iron Stove.
1 Tent.

*Supplies listed for one man's survival in the Yukon for one year.*

require for a year in the Klondike. This included eight 50 pound sacks of flour, 100 pounds of beans, 125 pounds of bacon, 25 pounds of rice, enough baking powder and yeast, evaporated milk, potatoes, and apples. Also required were the necessary cooking utensils, a variety of tools, and...only five bars of Castile soap![9] Hopefully, more soap was dispatched to the Yukon than the published list required.

There were some purchases that could only be considered foolish. A contraption referred to by hecklers as a 'Klondike bicycle' would prove worthless on the trail after only a few hundred yards. A device for automatic gold panning based on the gramophone principle, didn't find many buyers either.

Tappan Adney and his fellow argonauts boarded the *Islander* on Sunday expecting to leave on time. Unfortunately the ship was not ready until 5:15 p.m. the following day. The vessel, under the command of Captain John Irving,[10] arrived in Vancouver at dawn, Tuesday, August 17, where more men were ready to board, among them Robert and Arthur. Around them were crowded hundreds of other men, women, and children, all shouting "three cheers for the Yukon" as the vessel cast off its lines.

Also loading at that point was a detachment of Northwest Mounted Police under the command of Inspector Harper. Robert Harris became acquainted with the Mounties during the voyage and later provided valuable assistance on the way to

*Steamer moored at Vancouver Wharf.*

*Steamer Islander departs from Victoria, B.C. bound for Alaska. Filled to capacity with man, beast, and cargo.*

Dawson City. The *Islander* departed Vancouver four hours later. On board were 160 eager men, and a like number of horses and mules.

Although Tappan Adney was one of many who had elected to bring animals, he took particular note, as did Inspector Harper and his men, of a few like the Harris's who favored knockdown boats. These were made in sections that could be carried more easily and assembled later when needed. Robert and Arthur had selected for their use a style of boat known as a Peterborough canoe. They were familiar with these boats, first designed and made not far from Atherley. Larger than a typical canoe, bow and stern were both pointed so that in very heavy currents, like the swirling Miles Canyon rapids awaiting them, the vessel could be piloted from either end.

Weather along the way was clear and pleasant. Most on board likened the journey to a summer excursion as they passed the long hours retelling the many rumors overheard. They inspected each other's outfits and tended to the horses and mules. In between times, they played poker with an agreed upon five cent betting limit. There were boxing matches, songfests, anything to occupy their minds and distract them from their worries.

Canadians on board were concerned about the new policies of U.S. Customs agents in Dyea and Skagway. While their goods would arrive on American soil, they were all destined for Canadian territory. Would these supplies have to be placed in bond? Hardships would ensue if man and food supply were separated. An American official on board, Mr. P.A. Smith, reassured them of the cooperation existing between the respective governments. They need not worry about being separated from their goods. Everyone's goal was for men and supplies to reach their destination before winter set in.

On August 19, the *Islander* reached the town of Juneau, just one hundred miles short of Dyea. Juneau had been established in 1880 by Joe Juneau and Richard Harris (no relation).[11] Now, the town was a center for refueling steamers and outfitting gold prospectors. Taking on fuel, the *Islander* departed two hours later. If anyone had doubted the Arctic conditions ahead, the snow-capped mountains and glaciers now coming into view made a vivid impression.

After entering the Lynn Channel, the *Islander* encountered a thick bank of fog. Captain Irving elected to drop anchor for several hours and await clearing. Passengers had been asking many questions about landing procedures so this seemed a good opportunity for a meeting. Minutes were kept, including the following:

> *"At 11 p.m., August 19, 1897, a meeting was called to order by Mr. Genest...the following plan was decided upon: after arrival, a representative will go ashore, select a suitable place on the beach for landing cargo...this will be enclosed by ropes, and the enclosure policed...men armed with rifles... in shifts of eight hours...no goods can be removed except on written orders..."*

Someone asked during the meeting, "What is the penalty for theft in Skagway?" The reply: "They give a man twenty-four hours to leave; and if he doesn't, he's shot."

The *Islander* had been scheduled to arrive in Dyea, but the burden of transferring heavy cargo including horses and mules dictated a need for mooring as close to shore as possible. Captain Irving knew that Dyea faced shallow water and that Skagway's harbor was deeper, allowing ships with greater draft to come closer. As this first part of the long journey neared its end, every man stood at the rail, and through a faint haze they watched the shallow bay at Dyea pass by silently. Behind it in the distance they could see Chilkoot Pass. Nine miles further, they entered the harbor of Skagway. Behind it lay White Pass. Captain Irving gave the order to drop anchor. It was August 20, 1897. Now, the hardest part of the adventure was about to begin.

*A modern ship in the Lynn Channel approaches Skagway. Bow points to Chilkoot Pass and Dyea. To the right lies White Pass.*

# 5

# THE RACE FOR DAWSON CITY

***"A man can only have one kind of fever at a time, and they surely all have gold fever now."***

*A woman stampeder*

It was indeed a race for gold, but it was also a race against the clock and the approaching winter. Snow had fallen on the slopes above Skagway. That told Robert and Arthur how fast winter was approaching. They didn't want to waste a minute, but since they were part of the carefully planned Skagway landing party, they lent a hand to those using the White Pass route. Because they had outfitted themselves for water passage and wanted to reach the lakes more quickly, Robert and Arthur were taking the shorter Chilkoot Trail.

White Pass was considered more suitable for pack animals, a more gradual ascent, although fifteen miles longer than the Chilkoot. Its principle obstacle was a deep gorge with narrow paths cut into each side. The trail on the south side had already been claimed by railroad investors for grading. This left only the north side trail available for many hundreds of men and their animals. Twenty-five hundred prospectors had arrived during the week prior to the *Islander's* arrival. Among them were John Nordstrom and his partners. They had found White Pass Trail to be in very poor condition. The number of horses and mules brought up through the gorge had led to near breakdown.

Whilc the *Islander* was still being emptied of its cargo, word

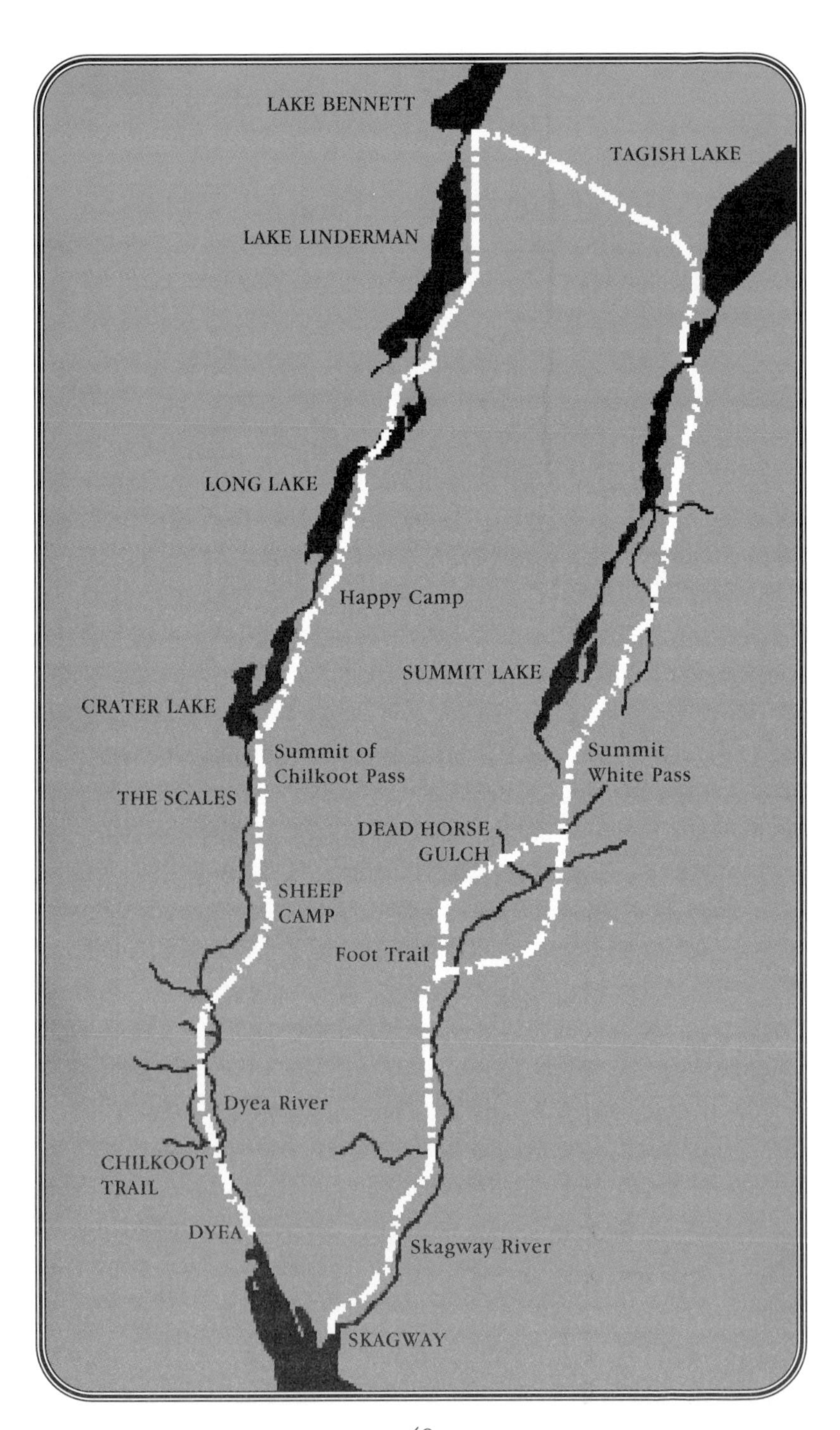
LAKE BENNETT
TAGISH LAKE
LAKE LINDERMAN
LONG LAKE
Happy Camp
SUMMIT LAKE
CRATER LAKE
Summit of
Chilkoot Pass
Summit
White Pass
THE SCALES
DEAD HORSE
GULCH
SHEEP
CAMP
Foot Trail
Dyea River
CHILKOOT
TRAIL
DYEA
Skagway River
SKAGWAY

*Supplies piled up at water's edge in Skagway, Alaska.*

arrived that White Pass was so backed up with animals and men that all effective movement forward had ceased. Journalist Tappan Adney interviewed a man just returned to Skagway who reported that a horse deliberately walked off the edge and fell to its death. "It looked to me like suicide", the man said. "I believe a horse will commit suicide and this terrain will make them do it."[1] In fact, 3,000 pack animals perished on that trail during 1897 and 1898.[2] So many fell to their death at one point that the site became known as 'dead horse gulch'. Eventually, horses and mules became a burden in Skagway. What had been purchased in Victoria for $27 a head, and on board ship for $100, were now on sale at $3 a head!

Taking note of these events, Robert and Arthur remained committed to making their way by boat. Others belatedly came to the same decision; knockdown boats were suddenly in demand. None were available for purchase at any price so they were sometimes appropriated. One man awoke in the morning to find his Peterborough canoe stolen and already on its way over one of the mountain passes.

Those members of the *Islander* party bound for the Chilkoot, Robert & Arthur among them, began transferring their goods by barge to Dyea or by portage along the connecting road. They were assisted by Indians who were plentiful and eager to help. Dyea had for many years served as an Indian

*Barges transferring supplies from steamers to Dyea Landing.*

settlement for three Tlingit[3] tribes: the Chilkats of the Lynn Channel, the Stikeens from the South, and the Chilkoots from mountains to the north. They were short, heavyset, powerful men capable of carrying two hundred pounds at a time over the pass when the price was right. And the price was an unprecedented 40 cents a pound.

Almost overnight, Dyea became a small city but its boomtown status was temporary. When railroad construction began, Dyea would be abandoned in favor of Skagway. Then, nearly every wood structure built during this frenetic summer of 1897 would be dismantled and the lumber taken to Skagway to be used again.

While the Harris's and other passengers were making their way from Skagway to Dyea, a heavy storm in the Lynn Channel was driving the tides to extraordinary heights. Beaches were flooded, submerging tons of supplies. Nearly all the Kodak film Adney brought to the Yukon was destroyed. Next day, he returned to Skagway where he posted a report to his New York editor and arranged with Wells Fargo for delivery of more film. It would eventually catch up with him in Dawson City.

As the Harris brothers prepared to leave Dyea, Jack London and his party were beyond Chilkoot Pass and camping at Lake Lindemann. Already, the hardships were more obvious. Many men were now rethinking their commitment and turning back. Hundreds would make that decision before reaching the summit, but not Robert and Arthur. Undeterred by the climate or terrain before them, they pushed on.

Between Dyea and Sheep Camp, the trail was an easy one for several miles, until it began ascending quickly into a narrow canyon. Next, the path climbed steeply to Sheep Camp, a way station in a half-mile wide valley, fourteen miles from Dyea and four miles short of the summit. Along the way, prospectors witnessed transport of some unusual items like a partially assembled 200 pound cooking range destined for a hotel under construction at Sheep Camp. Indian porters also carried loads of freshly caught salmon weighing 10-12 pounds each, for sale at 25 cents apiece.

Sheep Camp was little more than a village of tents and the last place this side of Chilkoot Pass where firewood was still available for cooking and warmth. At this point, men who were packing in by themselves returned to Dyea for another load. Without porters, it could take a man a week or more to lug his goods to Sheep Camp, then another week to haul them over the pass. Many individuals now reconsidered bringing pack animals any farther. Tappan Adney sold his horses at Sheep Camp for $50 a head. They could still be used for transporting cargo up from Dyea.

Already, Sheep Camp boasted a hotel constructed of wood,

not just a large tent. The proprietor, his wife, and children busily prepared several hundred meals each day. A pack train out of Dyea was entirely committed to keeping the hotel's kitchen supplied with food. A single meal of beans, tea, and bacon sold for the inflationary price of 75 cents, payable in advance. Sheep Camp became more than a convenient stop on the ascent to the pass. It grew into a community that absorbed great numbers of men incapable of going any farther. Some who needed to raise funds carried loads of supplies for other men. This led to competition between Indian porters and newcomers, or 'cheechakos', as white men were called by natives.

Beyond Sheep Camp, men in ever increasing numbers climbed the trail as it passed through a treacherous and even narrower canyon sliced between two mountain slopes covered with newly fallen snow. At the upper end of this canyon lay 'the scales', a place appropriately named, for it was where the Indians weighed miner's goods prior to taking them over Chilkoot Pass. The price for carrying prospector's supplies was renegotiated based on conditions and demand. Now, even before the heavy winter snowfall arrived, rates increased from 40 cents to $1.00 a pound.[4]

Rising steeply from 'the scales' was the infamous and feared grade to the Chilkoot Pass itself. No one who stood at its base would forget the never-ending chain of heavy-laden men. By the hundreds, they climbed and re-climbed that trail, gradually transferring tons of supplies into Canada. Once snow covered the grade, 1,500 steps, known as the 'golden stairs', would be chopped in the packed ice to aid a man's climb.[5] But in this final month of summer, the trail was muddy, wet, and slippery, yet still passable, especially to the native porters.

At the top, the trail leveled as it continued through the pass, then descended gently for the next three hundred feet to Crater Lake. This was the first of a chain of lakes that eventu-

*At 'the scales', looking up the 'golden staircase' to Chilkoot Pass.*

*Chilkoot Pass, looking back down at 'the scales' encampment.*

*Supplies stacked up at the top of Chilkoot Trail.*

ally drained into the Yukon River. Scattered along the way were makeshift tents and piles of supplies. With one member of a party standing guard, the rest descended again...and again, perhaps as far as Sheep Camp or Dyea to pick up the next load. Not all the men were adequately provisioned at this time, a situation that was troubling the Northwest Mounted Police.[6] By February, a man without sufficient goods to sustain himself for the winter would be refused admittance.

Crater Lake was a mile long and fed by the melting glaciers around its margins. Just beyond was Long Lake; where Robert and Arthur assembled their 26-foot boat. It was at this point in their journey that they recognized a financial opportunity. All about them were men who didn't have any idea what to do next after crossing Chilkoot Pass, or how to go about reaching their destinations. The brothers saw a way to earn money by assisting others for a fee. They could ferry goods to the end of the lake in their own boat, or pilot other men's boats, or show others how to build boats. But this opportunity to replenish

*Ferrying miner's supplies at Long Lake where Harris brothers first hired out to transfer cargo.*

spent resources also formed the basis of a dilemma. Each day, as the temperature dropped lower, they realized that there wasn't enough time to help everyone and still get to where they wanted to go. The time eventually came for them to move on. Beyond Long Lake was Lake Lindemann, then Lake Bennett where the Chilkoot and White Pass trails converged.

When John Nordstrom reached the top of White Pass, he realized that he needed to return to Skagway for more supplies before proceeding. This done, he met up with his party at Lake Bennett. Because of the abundance of timber, men without boats stopped there to build them. Nordstrom's partners were constructing a boat of sufficient size to carry four men with all their provisions. Having no further use for their horses, Nordstrom butchered them. At that moment horse steaks tasted pretty good, he later recorded. And, there was another bonus: it was now cold enough to freeze the hindquarters as a durable food source.[7]

By the time Robert and Arthur arrived at Lake Bennett, they could see what one observer later called the largest tent city in the world.[8] A sawmill was already up and working and trees

*World's largest tent city forms at Lake Bennett, winter of 1897-98.*

were rapidly disappearing. By winter's end, deforestation as far as the eye could see would be complete. Nails were in great demand, as were buckets of pitch to seal the newly constructed boats. Green wood and speed of construction guaranteed leakage, therefore slats were placed upon the cross-ribs to keep cargo dry. Boat builders were beginning to work under contract, charging $250 to $600 a boat. The largest were 35 feet in length but the most popular were 22-25 feet long, six to seven feet wide. Rigged with four pulling oars and a fifth steering oar astern, they were capable of carrying two or three tons. Enterprising owners recaptured spent funds by renting space on their boats. Many travelers chose this option rather than taking time to build or buy their own vessel.

Each day ten or more boats departed, usually to the accompaniment of loud shouts and an aerial fusillade from a bystander's revolver. Men by the dozens were sensing the greater hardship ahead. Many sold out, getting rid of all their goods for the best price, except for what they needed to get back to Dyea where they booked passage home on the first southbound steamer. Robert and Arthur forged ahead.

Leaving Lake Bennett, they negotiated the narrows at Caribou Crossing, paddling through shallow muddy waters into Tagish Lake. Two miles later, they spotted the red flag of Britain flying over a tent encampment under the command of James Godson of the Northwest Mounted Police. The Harris brothers had reached Canadian Customs. Earlier, when John Nordstrom and his party arrived, they had resented having to pay a 25 percent duty on their entire outfit, including the American-made boots on their feet! Assessors also realized that additional duties would be collectible from anyone taking gold out of the Yukon. Fortunately for the Harris brothers, Canadian citizens owed no duty on their personal goods.

The Mounties were increasingly concerned about the fate of men passing by. In the previous three years, the Yukon River had frozen over by the middle of October. Officers urged every man to hurry along the trail; before them lay the rest of Tagish Lake, then Lake Marsh, then 600 miles of the Yukon River itself before reaching Dawson City.

Miles Canyon above Whitehorse was a major obstacle. Barely one hundred feet wide, its walls were sheer rock. The swift current was filled with eddies that could spin a boat around in an instant. It was here that Robert and Arthur again elected to stop and pilot the boats of less experienced men. Charging $10-20 for each party according to a boat's size and load, they could make $300 in a good day. Among the men they assisted were some of the Mounties they had met on the *Islander*. For ten more days Robert and Arthur plied their trade on the rapids, earning a grand total of $2,475.

But day by day, warnings of the coming winter became more obvious. Ice crystals were forming each night at the edges of quiet pools near shore. During one of their passes through the rapids, they spotted a boat trapped on the rocks. Robert and Arthur tossed a line and managed to free the boat and its occupants. The owner, stricken with fear and fed up

*Piloting a newly-constructed craft through Miles Canyon. rapids.*

with the Yukon, agreed on the spot to sell the boat and its contents, provisions and rifles included, for $25! Now they each had a boat. Arthur took one and departed for Dawson City. Robert closed their piloting venture and soon followed in the other boat.

Ahead of them, Jack London and his party were entering Lake Labarge, actually a widening of the Yukon River. They reached Fort Selkirk on October 6. Three days later, they detoured up the Stewart River. A few miles upstream, London began staking claims on Henderson Creek but never found any gold there.

The Harris brothers were moving as fast as their oars and the current could propel them. Their only thoughts were about getting to Dawson City. Struggling for passage though mush ice and the gathering ice floes, they would be among the last to arrive by boat that season.

# 6

# LEAD PENCIL MINER

***"I'm a lead pencil miner, no pick and shovel for me."***
*Robert Harris, 1898*

A full year had passed since Joe Ladue staked his business empire on a meadow near the junction of the Klondike and Yukon Rivers. His vision of a thriving city serving the needs of a thousand prospectors had become a reality. In fact, Ladue's projected population already exceeded that number, and many more were on the way or planning their journey for the coming spring of 1898. He was exulted. However many came, Ladue would earn a fortune in his saloons, hotels, warehouses, and other enterprises. A significant portion of the gold found near Dawson City fell directly into his hands.

All 178 acres of the town site had been laid out by Mr. Ogilvie, the territorial surveyor, and Arthur Harper, Ladue's partner. By October 1897, more than three hundred commercial establishments had been constructed, each built of wood taken from surrounding forests and milled in Ladue's sawmill. The NWMP log compound quartered thirty constables. This three-sided complex facing the Yukon River included barracks, offices, a post office, and most important, a courtroom for adjudicating disputes. On a pole outside the main gate flew the flag of Great Britain.

A quarter mile away, an entire city block had been leased by the Alaska Commercial Company for a warehouse. A similar complex was erected in the next block by the competing North

*Front Street in Dawson City at the time of the Yukon Gold Rush.*

American Transportation and Trading Company. Numerous hotels bore predictable names like 'The Yukon', 'The Klondike', 'Dawson City', and 'Pioneer'. Saloons were plentiful too with colorful names like 'The Moosehorn', 'The Palace', 'The Dominion', and the 'M & M Saloon'. There was a brewery, an opera house, and several dance halls advertising scantily dressed girls. Two churches were completed, one of them Anglican and the other Catholic. More were under construction. And for the increasing number of sick and infirm, there was a small hospital, St. Mary's.

Joe Ladue's original cabin, the first structure in Dawson City, was now the town's principal bakery. Like his sawmill, it operated around the clock, as long as flour was available to bake. Other cabins became restaurants, mining brokerages, assay offices, and banks.

The town was set back sixty feet from the River's high water mark. Along the embankment lay hundreds of boats of all sizes, recently hauled from the water and secured for the long winter season. The last steamers were gone, departing the prior month for St. Michael.

Across the Klondike River was a Tlingit Indian settlement, dubbed 'Lousetown' by prospectors. Connected by a footbridge, the settlement drew a steady stream of visitors each night. Dawson City's respectable ladies had banished all prostitutes to Lousetown for the conduct of their professional duties.

Dodging the ice, Arthur passed by Lousetown before reaching the Dawson City embankment. Robert arrived a few days later, grateful to leave the icy currents. Their pockets filled with newly acquired cash, it was now time to begin looking for the gold nearly within their reach.

The best claims would be near the original strikes on the Bonanza and Eldorado Creeks. Robert and Arthur soon learned however that every claim on those tributaries had already been staked. Their only chance was to purchase a claim from an established holder. Providentially, this kind of opportunity presented itself very soon after they arrived.

A Mr. H. McCullough, who owned a half interest in a Bonanza Creek claim, needed to raise cash quickly. His claim, which he had purchased from Mr. Clement Blytheman, lay at #80 Below Discovery and had been registered just eleven days after Jim Skookum spotted his first gold nugget. The price agreed upon between Mr. McCullough and the Harris brothers was $3,000, $1,400 as an immediate down payment and the remaining $1,600 due and payable on July 1, 1898.[1] The deal was closed on October 22, 1897 and registered the next day. Robert and Arthur were very pleased with themselves; within a few days of arrival, they were part owners of a claim on the famed Bonanza Creek!

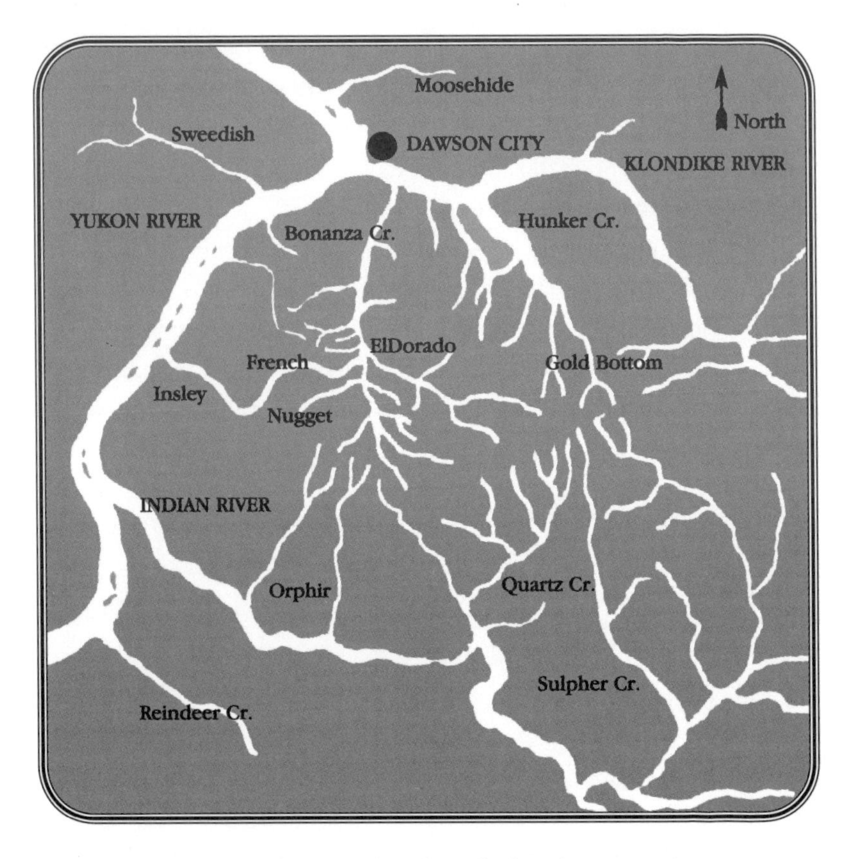


Just over a week later, fate handed Robert another oppotunity. It was even better than the first. Unforseen events led up to this moment. Back when the original claims along Bonanza and Eldorado Creeks were first staked, boundaries had been haphazardly established. Though the miners knew this, they were too busy to care. But, with the passage of time, border disagreements became more frequent and emotional disputes erupted. They appealed to William Ogilvie, the former government surveyor who had recently accepted the post of Gold Commissioner for the Yukon district.[2]

No miner dared challenge the authority or surveying skill of Ogilvie who had already resolved numerous disputes along many of the Yukon and Klondike tributaries. Now, in surveying the disputed area along Bonanza Creek, Ogilvie defined an

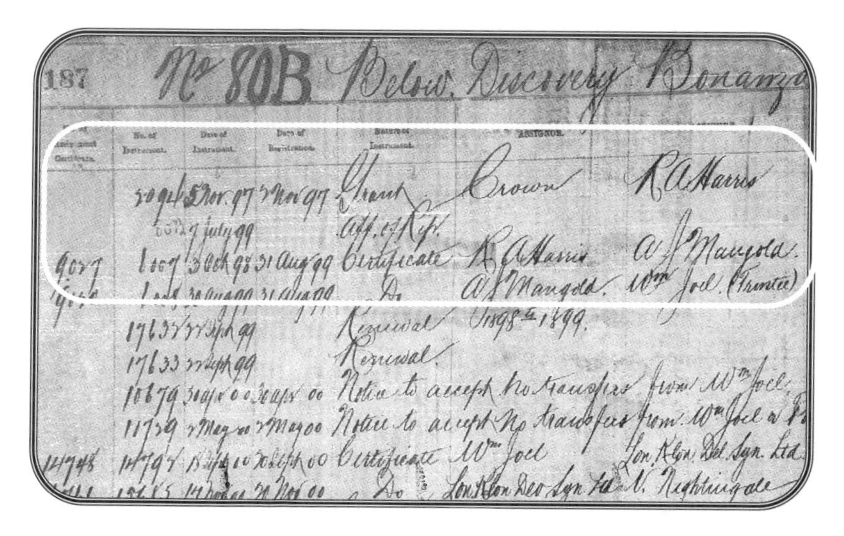


*A page from one of the 'creek books' showing #80b Below Discovery on Bonanza Creek.*

unclaimed wedge of land at the margin of Robert and Arthur's new claim. And Robert was standing right there when Ogilvie found it. Instantly he declared his claim, and left for Dawson City to file Application 2094 in the name of R. A. Harris. Only ten days after purchasing his first claim, he was awarded by the Crown at no cost an original fractional claim. It's designation was #80b Below Discovery, 56 feet in length and 200 feet in width on both sides of the creek.[3] Robert's declaration on the claim application read as follows:

> *"I have discovered therein a deposit of gold...*
> *to the best of my knowledge and belief,*
> *the first discoverer of said deposit."*

The bureaucratic language of this government form was based on the assumption the claimant had actually discovered gold. In truth, Robert had only acquired land which could be assumed to contain gold because of its location on Bonanza Creek.

Meanwhile, Arthur had been anxious to strike out and establish original claims on nearby but less famous creeks.

There was Moosehide Creek, a tributary of the Yukon, down river from Dawson City and now an easy walk over the ice. It was here that Arthur staked his claim for #63 Above Discovery.[4] He later rejoined Robert and learned about his brother's fortuitous second claim on Bonanza Creek. Together, they traveled back up the Yukon to Orphir Creek where Arthur staked a claim on a tributary joining that creek. This one he gave the family name, Harris Creek.[5]

At the same time, Robert found a gap in registered claims on Orphir Creek and staked #55 Above Discovery for himself.[6] He returned to Dawson City a week later to register that claim. The next day, Arthur registered both his Moosehide and Harris Creek claims. However, there is no evidence that they ever returned to or worked any of these claims.

The talk around town centered on food...would there be enough to sustain the population for an entire winter? The final shipment of the season, 125 tons of provisions, had been delivered by the *Portus B. Weare* September 28. Downriver at Circle City, miners had taken twenty tons of the provisions at gunpoint, fearing that too much would be delivered to Dawson City. But with river steamers not expected to resume deliveries until the following May, would it be enough?

On September 30, Captain Constantine had posted the following notices:

*The Canadian Government...finds that the stock on hand is not sufficient to meet the wants of people now in the district...can see but one way out of the difficulty...an immediate move downriver of all those who are now unsupplied to Fort Yukon where there is a large stock of provisions...it is absolutely hazardous to build hopes upon the arrival of other boats... in a few days the river will be closed."*

This did not much affect Robert and Arthur who had arrived with sufficient provisions for their own needs. But on the streets of Dawson City, Robert and Arthur overheard wild talk about seizing warehouses and allocating existing supplies to those who had been in the district the longest. There were only twenty Mounties to maintain order in the entire town. Nevertheless, the warehouses were guarded around the clock and no one ever challenged them.

*Claimsites along Bonanza Creek, summer of 1898.*

Bartering for food was commonplace because most men were unwilling to leave their claims even though winter conditions prevented panning or use of the sluicebox. One man with more than sufficient bacon or sugar could find another with an abundance of flour or beans.

Robert and Arthur, still not ready to engage in the labor of digging tunnels in search of paystreaks, went off into unexplored creeks to find their financial security. Specifically, they were looking for unclaimed gaps on less popular creeks for

*Miners digging a shaft during winter of 1898.*

later resale to newly arrived prospectors. Arthur found and registered original sites on Deadwood, Hunker,[7] Henderson,[8/9] and Reindeer Creeks.[10] Robert registered claims on Henderson[11] and Enslay[12] Creeks.

*Busiest place in Dawson City: the Gold Assayer's Office*

Then, two days before Christmas, something prompted Robert to begin selling claims...his own claims. This clearly signaled a remarkable change in attitude about his long-term commitment to the Yukon. Just as he and Arthur had been anxious to purchase a claim soon after arriving, latecomers would now pay even more for prime property. Robert sold his Henderson Creek claim for $275,[13] all of it profit since the land had been granted by the Crown. The Harris brothers began to calculate what their Bonanza Creek property might be worth when the river was flowing again.

For now, the full force of winter roared down upon Dawson City. On January 5, 1898, Arthur registered his last claim on a tributary of Reindeer Creek.[14] Robert's final purchases were two half-interests in claims on Hunker[15] and Dominion[16] Creeks. At this point, Robert and Arthur decided that they held enough land. Now they would wait for the hundreds, maybe thousands, of impatient buyers soon to pour into the region.

While Arthur saw no need to leave the Yukon, Robert desperately wanted to return to his family in Vancouver. He wondered how they were faring. Was the house providing Keturah with enough boarders? Missing his daughters, Robert took time to compose a poem for them:[17]

*I'm a lead pencil miner, no pick and shovel for me.*
*But of course I had to take them when I landed*
*at Dyea. Hazel and Nellie, Nina and Donnie, my*
*little daughters four, I left them home with mother,*
*fully in her care. I think I hear them saying*
*as through a telephone, Oh mother,can you*
*tell us when daddy's coming home?*
*Oh no I can't my darlings, but I hope it will*
*be soon, For with or without the money,*
*he's welcome home to stay.*

He mailed this poem in Dawson City as he had done earlier with many other letters. Little did he know that none would arrive until after he returned to Vancouver.

As the long winter days of inactivity wore on, Robert began preparing for a mid-winter journey out of the Yukon. Hiring an Indian guide, he packed enough food for the journey and they set out by dogsled, traveling overland to the lakes above the Yukon, eventually reaching Chilkoot Pass.

*Keturah.*

# 7

# KETURAH'S STORY

***"For the love of adventurous women, that we may love our adventurous selves."***

*Klondike Women, Melanie Mayer, 1989*

At the end of the nineteenth century, the exploits of men were commonly reported in public records, newspapers, and magazines. Only rarely did the achievements of women find any public recognition, perhaps because women were expected to support a man's vocation or restless yearnings. Thus, very little is recorded about Keturah Harris's role in her husband's Yukon adventure. But the oral history passed down by family members offers many clues.

Could anyone in Atherley, Ontario have imagined in advance that a conservative Whitney girl would pack up and cross a continent not just alone but pregnant and with three young girls in tow? Or could they imagine her running a boarding house in a frontier city? Certainly no one in Atherley anticipated any woman they knew taking such risks.

Keturah was the eighth of nine children, born on April 14, 1865 to Julia Ann Hager and Henry Whitney. She grew up on the family farm, Maple Grove, near Atherley on the northeastern shore of Lake Simcoe. Her grandfather and uncles had been Loyalists, fighting in the War of 1812 to save Canada for England. Her father had been a successful grain miller before moving to Atherley where he grew wheat and oats.

Keturah was a happy child, known in the family as 'Turie' or 'Tute', and was especially close to her older sister, Emily, and younger sister, Julia. As a young adult, she and Julia were courted by two sons of the Collingwood Harris family, Robert and Thomas, who lived in town. When her brother William joined Thomas and Arthur for a summer of railroad work high in the Canadian Rockies, Julia impatiently waited for Thomas to return. Meanwhile, Keturah was developing her own relationship with Robert who had remained behind to watch over the family interests.

Thomas returned at the end of summer and soon married Julia. Two years later Robert married Keturah on July 3, 1888. Her older sister Emily wrote in her diary of that day being lovely and special for her as well because she was the Maid of Honor. Forty guests were in attendance at the family home. After vows, there was a reception, she recorded, making special mention of the clear, fresh water served. A horse-drawn cab transported 'Turie' and 'Bert' to the railroad station in Orillia where they boarded the southbound Canadian Northern for a honeymoon in Toronto.

The newlyweds returned four days later, taking up residence in their new house at the narrows in Atherley. Emily delivered the many wedding presents upon their return. Later she wrote of the loneliness she felt after her two younger sisters left. But they visited often.

The following June, Keturah gave birth to a daughter, Hazel,[1] and was soon pregnant again. Despite all there was to do at home, and the steady growth of his family, Robert became restless to go west. Letters arrived from Arthur who had moved to Vancouver. In Ontario, the growing season was limited by the long winter season, and hotels along the lakeshore closed when the tourist trade left after the short summer season. In Vancouver, Arthur wrote, it was warm year round; the sounds and inlets never froze. And the vigorous new city was growing larger each and every day. Ida, their older sister, had already joined Arthur there. Together they wrote of the steady expansion of the salmon fishing trade. Reading this, Robert decided the time had come for him to make his move.

It is likely that Henry Whitney, Keturah's father, might have added his own encouragement. While visiting friends in central Canada, he observed that the soil was rich and plentiful, producing extraordinary grain yields. Upon his return to Ontario, Whitney told a newspaper reporter that he would encourage any young man with a grain farming interest to take up a homestead claim in the west. The cost of land there was ten dollars per acre. But with Arthur and Ida in Vancouver, Robert's only interest was the Pacific Coast.

He first visited Arthur and Ida in Vancouver in 1889. Periodic returns to Atherley produced more daughters: Nina in 1891,[2] and Donelda in 1893.[3] One of these visits coincided with a census in which Robert listed his occupation as laborer, but according to newspaper accounts of his travels westward, he declared that salmon fishing was his new trade.

Six years later, Keturah moved west and joined Robert in Vancouver. It was a most unusual sight in the 1890's: a 28-year old woman traveling cross country by rail. Years later she would recall three trips across the Rockies alone (meaning without Robert).

By the time Keturah and the children arrived, the Vancouver property that Robert had purchased for $300 in January, 1890 had been developed. Robert often listed himself as a carpenter, so it is likely that he had a hand in the construction of his house. Vancouver archives show maps depicting the outline of structures existing along each of the developed streets. The house at 624 Keefer Street was nearly twice the size of adjacent homes, consistent with family recollection that Keturah's responsibility, in addition to raising the girls, was to manage a boarding house. Keefer Street was an ideal location for families and boarders: close to a school and to the commercial district along nearby Water Street. The harbor and the CPR terminal were but a 15-20 minute walk away.

Keturah and daughters Hazel, Nina, and Donelda had come to Vancouver at a time of increasing prosperity for Robert and

Arthur. The Harris brothers, like other successful entrepreneurs, were by then participating in both the harvest and the processing of salmon. Salmon was selling for 1/2 cent a pound, enough in those days for the fisherman and the canner to profit handsomely.

A fourth daughter, Nellie, was born in September 1896.[4] Two thousand miles to the north, George Carmack and two Indian partners had just discovered their personal Eldorado. Another ten months passed before the Harris family began to understand the impact of that discovery on their own destiny. The news, when it arrived with the steamers from Alaska, amazed the family as much as everyone else. Robert and Arthur knew instantly they had to go. Like most wives of men bitten by the gold bug, Keturah never considered going to the Yukon with her husband. Mining camps were no place to raise children. Besides, a boarding house would provide income while Robert was gone. Demand for rooms was gaining momentum now that a rush for gold was underway.

There were a few women who embarked for the Yukon out of desperation, unable to support themselves or unwilling to face not knowing what their spouses were suffering, or perhaps enjoying. Ten percent of the stampeders who would venture north to the Klondike were women. Not all were performers or prostitutes. Some were creative businesswomen like Harriet Smith Pullen who arrived in Skagway with $7 in her purse and developed a profitable restaurant and bakery.[5]

In 1869, Ellen (Nellie) Cashman, born in County Cork, Ireland in 1845 and raised by her mother in Boston, arrived in San Francisco. Later she moved on to Virginia City, Nevada where she managed boarding houses and restaurants. When the gold and silver ran out, she joined a party of men traveling north to a gold camp in British Columbia. She found little gold. In 1897, news of the great Yukon discovery induced her to try again.[6]

Traveling only with her dog and a modest sized sled packed

with the barest necessities, she departed Victoria for Dyea in the winter of 1898. She talked her way past the Mounties even though she carried less than the required ton of provisions. Perhaps it was because of her diminutive stature, less than a hundred pounds. Dogged persistence carried her over the ice to Dawson City where she opened a restaurant.

At the same time Ellen Cashman was coming into the Yukon, Robert Harris was coming out. He found passable trails carved by previous overland winter travelers. Because of anticipated food shortages in Dawson City, Harris had agreed to carry 116 pounds of bear meat in trade for provisions he had buried in a cache 140 miles to the south.[7]

Robert was amazed by the scene he encountered at Lake Bennett. The surrounding forests had been stripped bare and replaced by enormous tent encampments. Several thousand treasure seekers were living there, all waiting for easier passage north once the ice melted. Here, food was more plentiful than in Dawson City because supplies were regularly coming over the passes. At Chilkoot Pass, he saw the Mounties inspecting all newcomers and collecting import duties around the clock.

Looking down at 'the scales', Robert could see newly constructed tramways as they lifted greater quantities of cargo in one hour than he and Arthur had carried on their backs in a full day. He descended the 'golden stairway' with far greater ease and less baggage than he'd carried in five months before. Sheep Camp too had developed considerably since he had last been there; it boasted several hotels and many more saloons. Construction, however, had begun to slow down because White Pass was taking a greater portion of the traffic. Dyea was now considerably smaller than he had remembered it. Skagway, on the other hand, was a boomtown - the principle port of entry to the entire region. Robert took little time to explore Skagway. With the steamer *Centennial* ready to depart for Vancouver, he grabbed the opportunity.

At home in Vancouver, Keturah and the children were anxious about Robert's welfare. Although Keturah was not given to excessive worry, there was ample reason for the family of any prospector to be concerned about conditions in the Yukon. She hadn't received any mail in months. Worse, newspapers were filled with tales of extraordinary hardships, reported by men who had chosen to turn back before reaching their goal.

As the earliest signs of spring arrived there was still no mail from Robert. Stoically, Keturah reminded herself that no news was probably good news. Had either brother turned back, she would surely know by now. Had one of them been injured - or died - the other would have let the family know. Yet...suppose both had lost their way or perished...? Keturah banished the thought from her mind.

On April 6, 1898, the *Centennial* tied up at Vancouver's wharf.[8] To Keturah's enormous relief, Robert was among the passengers. For daughters Hazel, Nina, Donelda, and Nellie, there was absolute joy in seeing their father safe at home. Later came the excitement that defied belief: the story of the treacherous journey, the gold-bearing claims on Bonanza Creek! What might this treasure afford them in the future?

*Photograph of the Harris daughters taken in Vancouver.*

# 8

# SECOND JOURNEY

***It ain't easy to find the stuff. Why half the time you run across it where it ain't.***

*Author unknown*

Newspaper editors were eager for any good news coming out of the Yukon. Robert was only too happy to grant an interview to a reporter for the *Vancouver Daily News-Advertiser*. The article appeared the day after *Centennial's* arrival under the headline, "A Vancouverite Returns - A Rosy Prospect For All."[1]

Robert was described as being well-known in Vancouver, "...a former manager and quarter-share owner of the English Bay Cannery". He described the journey northward with his brother, and the challenge of carrying their boat in three sections over the Chilkoot Trail. He told about earning $100 a day helping other prospectors pilot their boats. These unprecedented wages would be the basis of stories he would retell for the rest of his life.

Anticipating a significant appreciation since making his Bonanza Creek claim five months earlier, Robert projected a sale price of $4,000. He painted a verbal picture of the excitement generated by all newly-reported discoveries; men kept outfits by their side ready to join any stampede to a new location. Disappointment followed any announcement that did not 'pan out'. He and his brother had instinctively resisted a rush to Swedish Creek, which yielded nothing of value.

Detailing conditions and prices in Dawson City, Robert speculated about alternate routes he might take back to the gold country. He made clear that he was home only for a brief visit with his wife and children before returning to his claims. Robert predicted the resumption of a rush northward, offering his advice to the newspaper's readers: "A practical man can make more money by catering to the numerous requirements of the people than by washing the soil."

Although, Robert was going back, he did not plan to spend any more time in the Yukon than was absolutely necessary. Knowing that demand for gold-bearing claims would be even greater than before, he would return only to sell those claims he and his brother had already established. His belief was confirmed by the activity at Vancouver's wharves. The number of northbound steamers was already increasing, departing regularly with full loads of men and cargo.

Robert said farewell again to his family just two months after he had arrived home; it was now June of 1898. Reassured by Robert's safe return and his plans to be back out of the Yukon for good before year's end, Keturah and the girls packed for a visit with the family in Atherley. It was their second journey across the Rockies. Appearing in the local newspaper, *The Orillia Packet* on July 21, 1898, just a few days after her arrival was the following item: "Mrs. R.A. Harris and family, of Vancouver, B.C., are visiting her parents, Mr. and Mrs. H. Whitney."[2] This notice appeared on the social page without any mention of Robert's and Arthur's apparent success in the Yukon. But the family knew. They were spellbound by Keturah's telling of the adventure and especially about the gold claims.

It took less than a week for Robert to get back up to Skagway by steamer. A major part of the cargo being unloaded each day from ships docking at Skagway were steel rails needed for the railroad over White Pass. Construction had begun

May 27, 1898, although preparation of the rail bed had been proceeding for months.

This railroad was first conceived by three businessmen back in Victoria. They had applied for and received a charter soon after the Canadian government realized the magnitude of the gold discovery. But the Victoria partnership was not able to secure the necessary financial backing. In March of 1898, they sold their charter to the Close Brothers of London, who were well known for aggressive ventures of this kind. Moving quickly, they appointed an American, Samuel Graves, as President of a new corporation, The Pacific and Arctic Railway. He immediately incorporated in West Virginia, thus saving taxes and gaining a strategic right-of-way over the American portion of the route between Skagway and White Pass. The railroad would replace the original toll road owned by George Brackett, who received $100,000 for those rights.[3]

The population of Skagway had continued to balloon, making it even more chaotic than Robert remembered. Buildings facing the main street were now festooned with bunting and flags in anticipation of a rousing Fourth of July celebration. On that day, bands played, men marched, and women waved their handkerchiefs. Firecrackers exploded while revolvers were discharged into the sky above.

Serving as Grand Marshall of the Independence Day Parade was the infamous Jefferson Randolph Smith, popularly called 'Soapy Smith', although never to his face. As self-appointed dictator of Skagway, he was notorious for his scams and graft. He had personally arranged to be honored for "all that he had achieved in Skagway" by Alaska's Governor who just happened to be on hand that day. This was to be the personal validation for a con man who had first learned to exploit men's weaknesses during the Colorado rush for gold.

But Jefferson Randolph Smith would be dead four days later, shot through the heart by a man defending a prospector

Broadway, Skagway, Alaska, May 1898.

foolish enough to display his newly-found wealth. Like so many successful miners, this fellow was carrying the product of an entire winter's toil. Friends warned him not to flaunt his gold so readily, but he had failed to sense the impending danger. Smith's men, circulating through town searching for prey, soon found him! The man was deftly relieved of his $2,800.

Knowing just who the boss of Skagway was, police officers of course failed to act. And that is when a lone man named Frank Reid decided the crime was one too many for the town's inhabitants to endure. He confronted Smith on July 8, guns were drawn, shots were fired! Smith fell dead, like a sack of ore according to one observer.[4] Reid was severely injured. A railroad surgeon was summoned and he operated through the night. But Reid's wound proved fatal as well. As for Smith, remarkably his body lay where it fell, unclaimed for days. But on the day after the shooting, the entire town turned out to honor the town's champion; Reid's funeral was the largest in Skagway's history. As for the hapless miner, his gold was soon recovered and returned. But Smith's henchmen scattered soon after several were jailed.

No matter where Robert Harris might have been on his route that day or the next, he heard about Soapy Smith's demise, as did every other man in the territory. Spreading even faster than the original discovery of gold, news of Smith's death brought countless cheers. The entire Yukon, not just Skagway, was now a safer place.

Since most men now entered the Yukon by way of White Pass, Robert joined the crowds headed in that direction. There was less concern during the summer months about the adequacy of supplies brought in by men. Nonetheless, Canadian Customs was still collecting duties on the value of all goods, except those brought in by Canadian citizens like Robert, who was traveling light this time. He wasn't even required to show an outfit because he was a landholder returning to an established claim. At the same time, men were coming out of the Klondike with their hard-earned gold. Thus commerce was very active all along the trail.

The river had opened on May 8th and Robert found many more options for transport down the Yukon. Steamers left Whitehorse almost daily and he boarded one of them. Settlements along the great river had recently suffered floods when the ice floes had jammed, sending wild currents of water over the embankments. Dawson City itself, still unprotected by levees, had experienced its first serious flood that year. The streets that had been so carefully laid out the year before lay submerged below five feet of water for the entire month of May and well into June. As Robert neared Dawson City, he noticed that the footbridge serving Lousetown had been washed away. Later, he learned that people had lived for a time on their roofs, their boats tied nearby for trips to a store or the post office.

By July, an entire fleet of boats of every size was coursing up and down every navigable waterway. Traffic on the Yukon River facilitated commerce, bringing unimagined luxury to a city with unprecedented spending power. *Harper's Weekly* journalist, Tappan Adney, had remained for the winter and took note that the first eggs to reach Dawson City in May brought $18 a dozen, oranges and applies cost a dollar each, watermelons sold for $25 apiece! But after several steamers had come with greater quantities of fresh vegetables, fruit, milk and cases of eggs, the inflated prices fell quickly. Within a few weeks, eggs were selling for three dollars a dozen, still several hundred percent greater than their cost in Seattle and Vancouver.[5]

Robert soon found Arthur and gave him the news of the family and life in Vancouver. Traveling to Bonanza Creek, they witnessed the resumption of placer mining all along the shore. Those who dug shafts during the winter were now shoveling the excavated soil and gravel into sluiceboxes of every conceivable design. Flowing water was the essential factor that permitted separation of denser gold particles from the lighter soil and gravel.

Both Robert and Arthur, experienced hands now, were amazed to see men scrambling over hillsides above the creeks. One such slope was named 'Cheechako Hill' after the native word for newcomer, because everybody knew that gold could be found only in creek beds, never on the hillsides. Except this time everyone was wrong! The dirt on those slopes was found to be very rich in gold ore. Geologists can understand that valleys are the result of erosion by creeks over the passing of centuries. This particular valley had been filled with gold long before it was carved by creeks. The 'cheechakos' Robert and Arthur saw hadn't known this; arriving late they had no choice but to stake land high above the creek beds. Ironically, owners of these hillside claims, known as bench claims, would in time take greater wealth out of the land than did the original claim holders.

No claims had been filed, purchased, or sold by Arthur during the winter months. Nor is there any record of claim transfer by either one during the summer. Both brought dust and nuggets back with them, but we can only speculate about the source. Did either one work the claims themselves? Did they hire others to do it for them in return for a percentage of the take? Or was the gold they held received in exchange for services or the eventual sale of their claims? The record is not clear.

John Nordstrom in the meantime found himself tangled in a legal web. He had purchased his claim at a moment when another individual believed that he had staked and registered it properly. The matter would have to be resolved in court before the Gold Commissioner. Unknown to Nordstrom at the time was the fact that owners of the claim next to his had quietly discovered a paystreak that extended in the direction of his disputed claim. They offered him $30,000 for ownership of his claim but only after he won his lawsuit or settled with his adversary. This was almost more than the young Swedish immigrant could bear, more challenging even than what he had already endured in the Yukon.

Robert filed for sale of his own Bonanza Creek claim, #80b Below Discovery on September 29, 1898.[6] This was the property that had come from Mr. Ogilvie's re-survey of the original creek claims. Costing Robert nothing but the filing fee, it was sold for $8,000 to a Mr. A.J. Mangold who represented the Pioneer Trading Corporation, Ltd., a London-based syndicate.[7] Payment was not received at the time, and documents found do not indicate what down payment or promise was made. But the record does show that Robert's agreement with Pioneer would not be met easily or quickly.

No other transactions in the Klondike region were recorded that summer by Robert or Arthur. Presumably, they both revisited their claims to determine which were likely to yield gold based upon the activity of claims nearby. Robert's intention to return to Ontario was known to Arthur, who was willing to remain in the Yukon for a little longer. Arthur's plan was to eventually return to Vancouver, the city he would call home for the rest of his life.

As summer ended and the earliest signs of oncoming winter appeared, Robert made ready to leave Dawson City and the Yukon region for the last time. He secured passage on one of the last steamers to depart for Whitehorse before ice closed the Yukon River. Meanwhile, Arthur was seeking a buyer for the half interest in their first acquisition, the Bonanza Creek property #80 Below Discovery that he and Robert shared. A Mr. William Bradley agreed to purchase a one-quarter interest immediately and another quarter one year later.[8] The terms were $2,500 per quarter at the time of sale, the rest to be paid by August 1, 1900. Arthur Harris, as agent, signed for himself and his brother Robert on November 1, 1898. Thus, an initial investment of $3,000 yielded a return of $8,000, in addition to the value of any gold taken from the claim prior to its sale.

As Robert came back over White Pass for the last time, he could hear the frequent blasts of dynamite that created the ledge necessary for laying rails at that altitude. Workers had not yet reached the summit but they would by February of 1899. The rail line would be extended to Lake Bennett by July, 1899. It would be completed to Whitehorse in July, 1900. Robert boarded the train just beneath the summit and rode in comfort all the way down to Skagway. The train passed over immense wooden trestles linking the steep walls of the gorge above dead horse gulch.

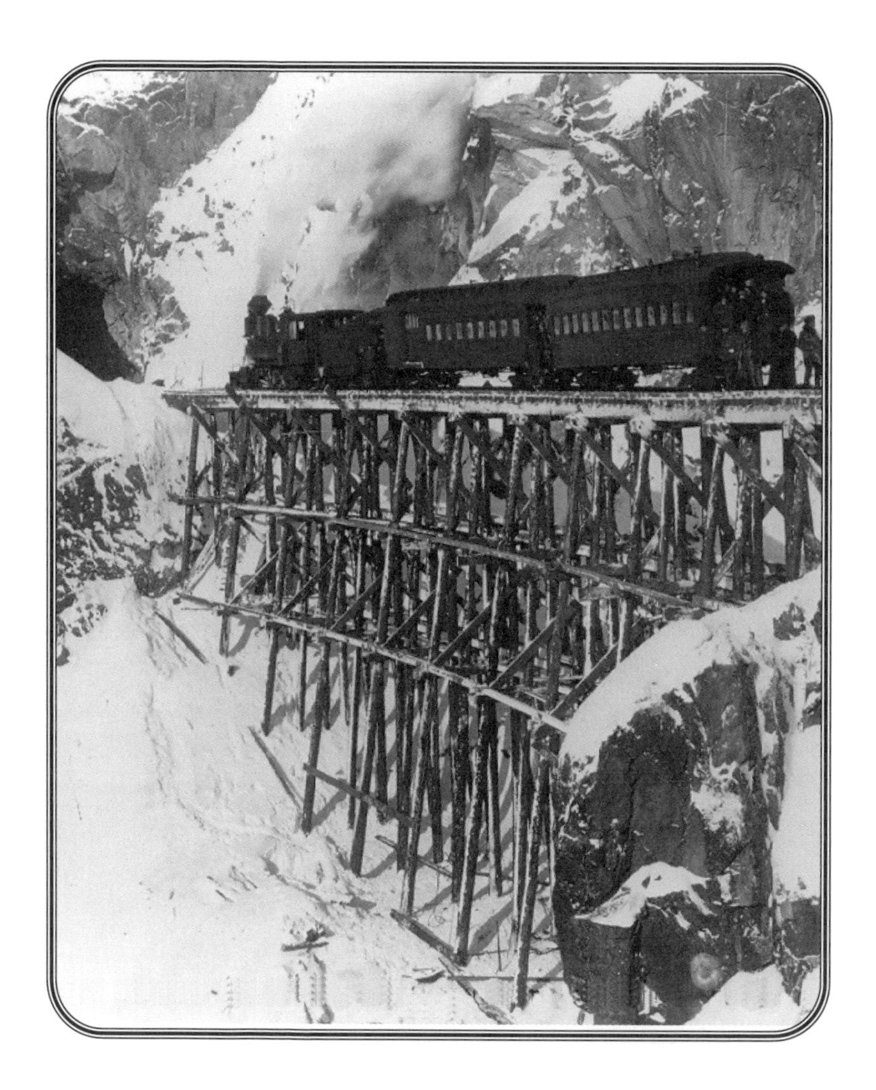

Railroad trestle spanning gorge above dead horse gulch with White Pass in the background, February 20, 1899.

Winter was closing in, and Skagway was filled with anxious prospectors waiting for steamers to carry them back to civilization. Robert carried with him $1,500 in gold and several thousand dollars more in the form of promissory notes offered as payment for claims sold. As he departed Skagway for Vancouver, the misty Lynn Channel receded behind him. In the distance were the mountainous outlines of the Chilkoot and White Passes he had come to know so intimately. We can only wonder what his thoughts and emotions were as he studied this scene for the last time.

*Robert A. Harris, age 36, photographed in Vancouver after returning from his successful Yukon journey.*

# 9

# BACK IN ATHERLEY

***It isn't the gold that I'm wanting so much as just finding the gold.***
*Robert Service*

Whether or not Keturah believed that Robert would actually return before the end of the year, she was thrilled to see him arrive in mid-October. Then when he announced that he would be taking the family back to Ontario and soon, she was doubly pleased. Her husband appeared to have satisfied his need to seek opportunity and adventure in the west. This meant turning his back on all of his interests in Vancouver: fishing, the cannery, the Keefer Street property with its boarding house income. No record indicates that he maintained any of these after returning to Atherley, although some time would pass for him to liquidate all of his assets.

Before the family could leave, however, Robert faced myriad tasks. Yet there was time for him to visit the Little Nugget Photo Shop where his four daughters had recently been photographed. Now he stood before the camera in a well-tailored suit and vest, his hair and beard neatly trimmed. He looked every bit the successful adventurer and businessman, his expression clearly reflecting a well deserved feeling of pride and accomplishment.

Meanwhile, Keturah went shopping for furniture and other household goods that might not be available back in Ontario. Then she packed all the family possessions for her fourth and final rail journey across the Rockies. Hazel, Nina, and Donelda bid farewell to their schoolmates at the Strathcona School.

Nellie, the youngest, knew nothing of Atherley and the family there, but joined with her sisters' excitement about returning home.

Family lore has Robert and Keturah leasing a private railroad car for the journey east. Whether true or merely apocryphal, Robert's new wealth allowed them to travel as comfortably as the Canadian Pacific Railway permitted. As their train climbed higher into the Selkirk Mountains and through Roger's Pass, snow covered the peaks around them. The reality of Ontario's winters must have reminded Keturah and the older girls of Vancouver's milder climate. But for Robert who recently experienced the bitterest of wilderness conditions, traveling through this mountain range in such comfort must have been a relief.

The transcontinental CPR linked up with track going south to Orillia where the Harris family disembarked. Two large wagons, each pulled by a team of horses were required to transport all of the family's belongings. Their arrival in Atherley must have been a sight, one that could have given rise to the notion that an entire railroad car had been required for their belongings. At the narrows, little had changed except that family members were a full decade older. It mattered little to anyone, especially Nellie, who had never seen them before. The following hours were joyous ones as parents, brothers, sisters, and cousins reacquainted themselves.

Local reporters were anxious to learn about Robert's adventure, and he was most willing to oblige their requests. The first interview was summarized in *The Orillia Packet* on December 1, 1898 under the lead, "Mr. R. A. Harris Comes Home With a Nice Pile!" Either the reporter forced the key question or Robert was willing that everyone should know he had returned with $1,500 as well as another $31,000 from the sale of his gold claims. Indeed, it was a nice pile. However, Robert was counting his chickens before they were hatched because many transactions were not yet completed. Still, he was clearly proud of his Yukon dealings and not about to miss the opportunity to let his neighbors know how clever he was. What story in any local newspapers in 1898 could have been more compelling?

Robert went on to tell how he had acquired the claim on Bonanza Creek. Defining himself as one of the first Canadians who had made any money in the Yukon, he may have been stretching the truth a bit. But it didn't sound as far-fetched when he also told of making $2,475 in ten days ferrying parties down the Miles Canyon rapids.

Dawson City, he explained, was a place with ten thousand inhabitants, most carrying their own miniature scales to measure gold dust, the prevailing currency of the region. Much of his business, he related, had been conducted through Dawson City's branch of the Bank of British North America. When questioned about starvation, Robert denied that it had been a problem for him. The people there he considered rough but orderly, with surprisingly little drunkenness or brawling. But, Robert added, the Yukon was no place for women.

As for his future plans, Robert told the reporter that he had not yet decided if he would return to Dawson City. For the time being, he would remain in Atherley while his brother continued looking after his business interests in the Yukon. Indeed, once the excitement and glory of what he'd done began to pass, Robert had to decide what he would do with the rest of his life. Not inclined towards farming, he turned to one of his father's favorite pursuits, hotelkeeping.

The elder Harris had built the *Hotel Malakoff* in the town for which he had been named, Collingwood, Ontario.[1] After moving to the narrows, Robert's father had recognized an opportunity to attract fellow fishermen to the lakes region. Collingwood constructed his second hotel, *The American*, not far from the narrows bridge he had built nearly a decade before. *The American* opened for guests on May 13, 1870, with boats and canoes available for hire by patrons. There was horseback riding too, and in the bar, "first class liquors and cigars" were served.[2]

When Collingwood Harris died fifteen years later, *The American Hotel* was purchased by Captain John McDougal. This colorful Scotsman, who according to one observer, had arrived in Atherley drunk, eventually joined the Harris family by marrying Robert's sister, Emma.

Ever since Captain McDougal's arrival, he'd been known as

a 'remittance man'. The term was a scornful one, acknowledging someone who received a regular subsistence check from a distant family, usually because of some wrongdoing elsewhere. Indeed, the rumor circulated that McDougal's family in Scotland was fabulously rich, that he had somehow disgraced them, and the Captain was being paid to stay away. But the evidence shows that he was retired from the Scottish Brigade of the British Royal Army, having served in Egypt and India.[3] His remittance was very likely nothing more scandalous than a pension check.

In any event, the Scot received funds regularly without appearing to work for them, making him a romantic figure in that day and age. The ladies considered him an aristocrat, a 'handsome Lockinvar' according to one admirer. But it was Emma Harris who captured the Captain's affections.

When McDougal purchased *The American Hotel*, he changed its name to *Peninsula Cottage*. There he celebrated his fortieth birthday. *The Orillia Daily Times* reported that 200 guests attended the gala event, including the leading citizens of surrounding counties and a few from Toronto.[4] Everyone enjoyed an afternoon and evening of fine food and entertainment on the tennis lawn, decorated with Chinese lanterns. A band played in concert. Then, at the conclusion of flowery speeches, dancing commenced.

There were some who believed that McDougal had bankrolled Robert's journey to the Yukon. Evidence for this is lacking. However, the Captain displayed immense interest in the details and outcome of Robert's quest for gold. And Robert undoubtedly consulted with McDougal before he committed himself to the hotel business. By that time, Robert and everyone else knew that the Captain had put his *Peninsula Cottage* up for sale.

Robert had been expanding his local land holdings even while living in Vancouver. Back in 1891, he had purchased his first lot at the narrows for $760. Whether this was done before he departed for Vancouver, or during one of his periodic returns to Atherley is not known. On May 11, 1899, he purchased seven additional lots totaling more than two acres. Included in that purchase was McDougal's *Peninsula Cottage*.

Robert was in touch with current financial conditions. As the nineteenth century drew to a close, the opportunities for attracting summer guests to Ontario's lakes was greater than at any time during the previous decade. Two rapidly growing metropolitan centers, Toronto and Buffalo, were a reasonable train ride or automobile drive away. The steel mills of Buffalo's lakeshore polluted the air, making summer respites at Canadian lakeside resorts a favorite choice for the well-to-do.

Nevertheless, thoughts of transactions in the west must have crossed his mind from time to time. Letters from Arthur described continuing efforts to liquidate all of their holdings in the Yukon. His brother visited him in Atherley in November 1899. Then Robert joined Arthur one last time in April, 1900, visiting Vancouver but going no farther. Their final joint transaction in the Yukon would be completed only after filing suit against the Pioneer Trading Company in December 1901. One month later, *The Orillia Packet* announced that Robert and Arthur had won their lawsuit, a settlement having been quickly agreed to by Pioneer. The Harris brothers were no longer landholders in the Yukon.

Arthur returned once more to Atherley because he was courting a local girl, Kate Marshall. They were married at St. Paul's Church in Vancouver on January 21, 1904.[5]

Robert in the meantime had focused his entire energy on his expanding business. He had changed the hotel's name to *Trondyke Beach House*. Robert was probably the only person in the Province of Ontario aware of the original native word for Klondike. Family members recall that local recognition of Trondyke was rather limited, so the hotel was later called the *Klondike*. Whatever the name, this first hotel simply did not complete his vision. He wanted to build a larger, more modern facility at the narrows. News of the project was first reported in *The Orillia Packet:* Harris would build a three story brick structure designed for the special needs of tourists visiting nearby lakes during the summer season. Its Grand Opening Celebration was announced in the local paper May 30, 1901.[6] It would be known as the *Peninsula House*. Its elegant brochure declared that "gentlemen will find it a most convenient place to send their families for the summer...not too far away from the business centres". Telephone connections were the finest available. Mail was delivered several times each day. The proprietor wanted to attract leaders of the business community. The rooms were large and airy, fitted with electric lights, good mattresses, and fine linen. Improvements had been made to the grounds. Pictures of the waterfront and nearby golf course graced the brochure.

*Peninsula House* could accommodate 75 guests at one time. The tariff was $10-12 per person per week, not inexpensive for that day. Every member of the family helped out during the peak summer season. A private carriage shuttled guests from the train station in nearby Orillia.

Indeed, Robert Harris was the proud owner of Atherley's newest and finest hotel. It must have been a high point in his life, the culmination of his ten year quest for success and prosperity which he found first in Vancouver, then in the Yukon, now in his own hometown.

# 10

# AFTERMATH

***You've read of the trail of '98***
***but its woe no man may tell.***
*Robert Service*

What became of the participants in this adventure, the territory they invaded, the cities and nations they returned to?

Arthur Harper didn't live long enough to enjoy any benefits of the gold rush; he died of tuberculosis in 1897 as the stampede to Dawson City gained momentum. Jack McQuesten, once the "King of Circle City, Alaska", chose not to move his business to Dawson City. Instead, he left the region, moving to San Francisco where he lived an easier life until his death in 1909. His partner, Joe Ladue, sold most of his lots within the 160 acre claim that became Dawson City. Prices went through the roof as more gold seekers arrived and he became a very wealthy man. But the demands of hard living in the Yukon had taken their toll. Ladue also died of tuberculosis, in 1901, after returning to his home in Plattsburgh, New York.[1]

George Carmack left his Indian wife, Kate, then remarried and moved to Seattle. From there he continued to search for another Bonanza Creek, first in the nearby Cascade Mountains and later in California. He never found it. Carmack died of pneumonia in Vancouver on June 5, 1922. His estate of $150,000 was fought over by family members and their respective attorneys for many years.[2] The lives of Skookum Jim and Tagish Charlie were essentially ruined by the gold rush. Never accepted by white men and unable to resume their cultural traditions, they spent the rest of their years wandering the Yukon's tributaries, never again finding gold. Charlie drowned near Lake Bennett in 1909; Jim died of kidney failure in 1916.[3]

Robert Henderson continued to live a life of endless bad luck. The Canadian Government awarded him a lifetime pension, preferring to acknowledge a Canadian rather than an American for the greatest discovery of gold within its boundaries. He lived for decades on his pension, but never found his own personal Eldorado.[4]

William Ogilvie remained the most respected public servant in the region and served as Commissioner of the Yukon Gold District from 1898-1901. Government pay and pension did not provide much; he died a poor man in 1912.[5]

During October and November of 1897, Jack London had staked several claims on Stewart River tributaries before Robert and Arthur reached the Klondike. Traveling overland between his claims and Dawson City several times that winter, London neglected his diet. By late spring, there were obvious signs of scurvy. Discouraged and severely ill, he left for Dawson City to await the first steamer heading for civilization. Departing on June 8, 1898, he sailed downriver to St. Michael, then on to San Francisco with little gold to show for his effort. However, a vivid recollection of his Yukon experiences later made him a rich man and favorite author, then as now. His first book, *The Son of Wolf*, was published in 1900. His most enduring work, *Call of the Wild*, followed in 1903. *Sea Wolf* and *White Fang* came later.[6]

John Nordstrom eventually agreed to settle the lawsuit affecting his disputed claim, paid off his expenses, and returned to Seattle with $13,000 in his pocket. There, Nordstrom entered the retail shoe business. Sometime later, he discovered that the buyers of his claim had taken out two million dollars in gold. However, the business he established became a gold mine of its own kind. He and his descendents founded one of the most successful clothing store chains in America. John Nordstrom died in his ninety-first year, outliving most of his fellow prospectors.[7]

New York journalist, Tappan Adney, continued filing reports from the Klondike until September of 1898, when he took a riverboat down the Yukon to St. Michael. He transferred to an ocean steamer bound for Seattle, bringing to an end his sixteen-month adventure. In 1900, he returned to Alaska to report on a new gold rush in Nome. After retiring, he wrote *The Klondike Stampede*,[8] a fascinating book widely considered to be the best first-person account of those historic years. He died in 1950. The Harris' descendents would like to believe that Adney's description of "two partners who stopped...to pilot boats through the canyon...and made enough to buy an interest in a Bonanza Creek claim", actually referred to Robert and Arthur Harris, fellow travelers on the *Islander* in August of 1897.[9]

The steamer *Islander* continued on the Victoria to Skagway route until the night of August 15, 1901. In the Lynn Channel just off Juneau, it struck a submerged iceberg and "sank in fifteen minutes" according to newspaper accounts of the tragedy. The captain, 16 hands, and 23 of 165 passengers were lost. The persistent belief that the vessel carried "a quarter million in gold" led to multiple attempts to raise the vessel. When this goal was finally achieved in 1934, there wasn't enough gold on board to pay for the salvage costs.[10]

Also participating in the Yukon adventure were more than one hundred thousand fellow aspirants to gold discovery. Only forty thousand actually reached the Klondike, and very few of those returned with resources of any value.

The cities of Seattle, Vancouver, and Victoria retained most of their inflated population, even after the Yukon stampede had waned. Victoria remained the capital of British Columbia,

Vancouver soon became the largest city in that Province, and Juneau became the capital of Alaska. Skagway remained a significant port of entry throughout the first half of the twentieth century, especially during the second world war when supplies and manpower for the Alaska Highway, a defense project of the U.S. Army, were carried in by the rolling stock of the White Pass and Yukon Route.

Fort Reliance, Fort Nelson, Fortymile, Fort Selkirk, and Circle City essentially disappeared after the gold rush, except for traces of turn-of-the-century cabins and a handful of more recent structures. Dawson City continued to enjoy an existence long after the frenzy of the stampede. Within a few years, nearly all of the individual claim holders would sell out to large business interests. The syndicates brought dredges and other technology into the region to rework the soil along Bonanza, Eldorado, and several other productive creeks. When Robert Harris left, the gold yield between 1885 and 1898 was $15 million. Over the next five years, gold production averaged $16 million a year. Cumulative production between 1885 and 1912 totaled $175 million![11]

The Canadian and American economies further stabilized during the years of the Yukon stampede, and continued to improve throughout the next decade. The United States treasury was never fully depleted of its gold stores. Instead, the nation's reserve grew in part because of the ore brought back from the Klondike by Americans. The gold standard was to be sustained well into the twentieth century.

Arthur Harris returned to Dawson City in 1902. We don't know what business he transacted at that time. Records show that he departed in September of that year, listing Victoria, B.C. as his forwarding address.[12] After bringing Kate to Vancouver in 1904, he served as captain of a small steamer, the *Terra Nova*, which was owned by a salmon cannery. Later, he attained the rank of Commodore for the British Columbia

Packet Company. His death at the age of 55 was announced in *The Orillia Packet* on October 28, 1915.

Robert Harris had spared no expense in building a thoroughly modern hotel. But there is clear indication that the fortune he brought back from the Yukon was insufficient to cover his costs. Even before the *Peninsula House* opened, he placed two mortgages on the property totaling $1,500.[13] Mr. Thomas Venner of Orillia held those mortgages during the first two years of the hotel's operation and then loaned Robert another $500. Public records show loans to other businessmen as well. Some loans were for uneven amounts like $565.35, suggesting that local merchants were gaining payment for goods by placing encumbrances on the property.

Robert knew that his business was failing. He turned to Captain McDougal who in 1905 agreed to pay off nearly $3,000 worth of locally held notes, including Mr. Venner's original investment. When the Captain's wife, Robert's sister Emma, died in 1907, Robert dispatched his eldest daughter Hazel to take care of McDougal's house for him. But it wasn't enough to keep him in Atherley, and the Captain soon moved to Toronto and remarried. He died in 1911.[14] It is not likely that his loans to Robert and Keturah were ever repaid.

Desperate for any resource that might improve his financial position, Robert learned of a gold discovery in Northern Ontario. In fact, The Gold Rose Mining Co. of Sudbury, Ontario, sought him out as a consultant. Specifically, the owners asked him to look at the claim site and to offer technical advice. Excited by all the attention, Robert accepted their invitation. He was eager for another opportunity to cash in, and he believed the Sudbury strike to be very promising. But he was wrong. Robert also thought that he had been granted exclusive rights to sell Gold Rose shares in Atherley. It turned out that the company had granted those rights to many others as well, and failed to pay Robert the promised 25% commission.

Gaining nothing from his efforts, Robert sued Gold Rose but without success. The family eventually received certificates for several hundred shares. By that time of course, they were worthless.

Circumstances at home were deteriorating quickly. Bills for basic maintenance were not being paid. Mechanic's liens were being placed on the property for amounts as small as $15.00, and as large as hundreds of dollars. In September, 1912, all the existing liens on the property were consolidated into a single mortgage for $4,500.

On February 5, 1914, tragedy struck the entire Harris family. As Robert's youngest children walked home from school, they could see black smoke issuing from their hotel. In terror, they ran the rest of the way home and saw their father carrying furniture out the front door as flames enveloped the second and third floors. The *Orillia Weekly Times* reported on February 12 that Robert's hotel had "taken fire and burned to the ground". Every effort was made by the Orillia Fire Brigade but the building could not be saved. The hotel's value was placed at $10,000 and the furnishings at $1,500. But there was only $5,000 worth of insurance on the entire establishment.

Robert was able to rebuild a house for his family from unburned lumber retrieved from the site of the fire. But he never recovered emotionally from the tragedy. Never again would he commit himself to any major effort. His daughter, Gretta, recalled that "Papa owed money all over town". Mancel, his son, was embarrassed to learn that a prized gold Waltham watch had been pawned.

Robert took ill and became too weak even to go fishing on his own. Mancel recalled carrying him out onto the ice in early December of 1922 so he could fish one last time. He died on December 11th at the age of 60. The cause of death was listed as kidney failure.

What of the Harris women? Ida Harris did not remain in

Vancouver after her two brothers left for the Yukon in 1897. She returned to Atherley and married Jake Gaudaur, famous in his community as an Olympic sculler. The Gaudaur family was not exactly pleased by this marital union; the two families had competed for years over the narrows bridge tending contract. Ida died in 1911.

Keturah resolutely maintained the Harris tradition for hospitality, very slowly piecing together a summer tourist trade,

Keturah Harris at age 85, sitting in front of her house at the narrows in 1950.

though on a much smaller scale than before the fire. Nearly all the Harris land at the narrows had been claimed by creditors but gradually, lot by lot, most of it was repurchased by the Harris children. Keturah died on June 6, 1952 at the age of 87. Cottages were built nearby her house where summer guests returned year after year until 1987.

Hazel, Nellie and Nina never married. Donelda[15] and Dorothy[16] married local men, and Gretta[17] married a man from Buffalo. Of the three brothers, Oswald[18] was the only one to marry.

Mancel continued to live and work at the narrows, operating the Harris Boat Livery. He outlived all of his brothers and sisters. During his life, he wrote and published two books of poetry entitled *"Poems"* and "*Poems of Faith*". When he died in 1987, the value of the Harris property was approximately $600,000, about the same as the proceeds of Robert Harris's Yukon adventure.[19]

*At the narrows, near Orillia, Ontario, today.*

# EPILOGUE

# IN THE YUKON TODAY

Anyone with a desire to follow the trail of the Klondike stampede today will experience a far easier journey than did the heroes of this story. Evidence of their experiences can still be found in abundance, especially during the years of centennial observance, 1996-1999.

Harbor improvements in Seattle have replaced Schwabacher's Wharf where crewmen on the steamer *Portland* threw lines to the waiting crowd. Displays and video presentations in the National Park Service Klondike Gold Rush Museum[1] provide visitors with a decidedly American perspective on Seattle's role in the Yukon stampede. *The Seattle-Post Intelligencer*,[2] whose earlier editor dispatched reporters to meet the steamer *Portland*, has published continuously from 1863 to this day. The University of Washington Library[3] in Seattle maintains the largest collection of original photographs from that era, several reproduced in this book by permission of the special collections librarian.

Victoria, B.C., located at the southern end of Vancouver Island, still has its picturesque harbor and the majestic Empress Hotel,[4] originally built by the Canadian Pacific Railway. The natural harbor's perimeter approximates the dimensions that existed in 1897. Victoria's Maritime Museum[5] maintains a collection of memorabilia from steamers like the *Islander* which departed frequently for Alaska once the discovery of gold was known. But the actual wharf these vessels departed from has long since been replaced.

The Canadian Pacific Railway remains a prosperous and diversified enterprise. However, passenger traffic on its rail lines is a small fraction of what it was in 1897. Visitors to Calgary, Alberta, and nearby Banff can ride on special excursion trains of the CPR or drive a major divided highway up the Bow River Valley to Kicking Horse Pass. Instead of witnessing the rapid and dangerous descent of westbound trains of yesteryear, they can watch cargo trains descending gradually through modern spiral tunnels based on Swiss technology. Trains bound west to Vancouver continue over Rogers Pass, itself made safer and more stable by reconstruction long after the pioneering work of the 1880's.

As CPR managing director, William Van Horne, predicted when he selected the town of Granville as terminus for the railroad, the city of Vancouver has experienced enormous growth. Visitors can still enjoy the oldest part of the city, known locally as 'Gastown'.[6] Walking down Water Street, today's visitors see several buildings that existed in 1897, including the original Hudson Bay Company Building.[7] The CPR replaced its original terminal building in 1912. The new terminal was recently renovated to serve as a tourist information center.

The 600 block of Keefer Street, site of the Harris home and boarding house, now borders Vancouver's Chinatown. All of those original homes are gone. A modern condominium development stands where we believe number 624 probably existed in 1897. But the original Strathcona School still stands in the next block, surrounded by more recent additions to the original brick structure.

Three additional resources for Yukon Gold Rush information are the Vancouver Public Library,[8] The British Columbia Archives,[9] and the Vancouver Maritime Museum.[10] The latter facility faces English Bay where numerous canneries processed salmon one hundred years ago.

The principal routes of travel to the Yukon have been altered significantly since the advent of commercial air travel. But remnants of the rail and water routes used by gold stampeders can still be found. Those going to Skagway from San Francisco, Seattle, or Vancouver can select from numerous cruise lines but must allow a few more days time for passage than was considered necessary in 1897. Travelers looking for economy can reserve space months in advance on the Alaska Marine Highway,[11] Alaska's superb ferry system, which departs Bellingham, Washington with stops in Vancouver, Prince Rupert, Ketchikan, Wrangell, Sitka, Juneau, Haines, and Skagway. There are special cruises to the Bering Sea, but St. Michael no longer exists. Nor is there scheduled service up the Yukon River to Dawson City, or even downriver from Whitehorse.

The number of ships cruising Alaska waters far exceeds the number of steamships operating at the time of the gold rush. There were 371 arrivals in Skagway during 1997, bringing more than 600,000 cruise passengers to walk the streets of Skagway during each summer season. Many of them will ride the restored railroad to White Pass, and some will board special buses for a side trip to Dawson City.

Skagway is carefully restored and maintained by the National Park Service. The harbor has been dredged and the wharf modernized to accommodate as many as seven cruise ships at one time. Between two and five may arrive on any given summer day. Many of the original structures from the time of the stampede can still be seen. The home of Captain Billy Moore, founder of Skagway, is now a museum. The picturesque Golden North Hotel still accepts overnight guests.[12] The fully restored White Pass and Yukon Route carries trains from Skagway several times each day during the summer season. First opened for operation on May 28, 1898, it remains the northernmost railroad operated in the Western Hemisphere. Today, passengers can ride in comfort up the gorge, past dead horse gulch, and across a modern steel replacement of the original wood trestle bridge. After reaching

*Modern Skagway street scene.*

*Skagway and its harbor today.*

the summit of White Pass and the Canadian border, WP & YR continues to Lake Bennett but no farther. However, bus connections to Whitehorse and Dawson City are available.

A nine mile graded road skirts the Lynn Channel toward the former town site of Dyea. Only the foundations of a handful of cabins can be found there. Nearby, the trailhead for Chilkoot Pass is well marked with appropriate warnings for hikers: pre-register with Canadian Customs, anticipate sudden changes in weather conditions, and don't expect to complete the climb in

*Chilkoot Trail Head.*

one day or even two. Three days is considered a minimum for treks over the pass and as far as Lake Bennett where the railroad can return hikers to Skagway.[13] With more than 3,000 hikers climbing or descending Chilkoot Trail each summer, park rangers have established a strictly enforced daily quota of fifty.

A modern highway also climbs out of Skagway, over White Pass to Lake Bennett, and then on to Whitehorse in the Yukon Territory. A modern hydroelectric facility has replaced the Miles Canyon rapids, just short of Whitehorse. The city is now the Capital of the Yukon Territory. The fully restored *SS Klondike* rests at its moorings on the Yukon River there. Other historical sites in that community include the Macbride Museum[14] which faces the river, and the Yukon Archives,[15] now located in a new expanded facility outside of town.

The highway from Whitehorse to Dawson City does not follow the course of the river. But Air North,[16] headquartered in Whitehorse, currently flies modern Hawker-Sidley turboprop equipment to Dawson City, on a route that closely parallels the river. The flight takes 65 minutes. In 1994, they were flying fully restored DC-3's and it took us longer. To travel by boat from Whitehorse to Dawson City requires a charter. Yukon historian Pierre Berton wrote of just such an adventure with his family in 1973.[17]

Dawson City is a mecca for anyone seeking Yukon history and nostalgia. Today, its principle industry is tourism, but there are prophets who envision another gold-based economy there, perhaps very soon. Its year-round population is 2,000. In the summer, this number doubles in order to service the influx of tourists, currently 65,000 each season. A mixture of modern and turn-of-the-century structures now reside on the meadow claimed by Joe Ladue.

An enormous levee now surrounds the town site making floods a rare event. A partially restored river steamer, *SS Keno*, rests on that levee, not far from the Canadian Commerce Bank Building where the poet, Robert Service, was once employed as a teller.[18] Several other structures built at the time of the stampede are still carefully preserved; among them the Yukon Hotel, the NWMP jail, and the British North American Bank where both Arthur and Robert Harris transacted some of their business. The Dawson City Museum and Historical Society[19] now occupies the former Territorial Administration Building on Fifth Avenue.

For a customized tour of the gold fields, we commissioned David Taylor, known locally as 'Buffalo'. A colorful guide with an encyclopedic reservoir of gold rush history and Klondike lore, he has operated Gold City Tours (across from the *SS Keno*) since 1983. No visitor should pass up the opportunity to spend at least half a day with him.[20]

As we departed Dawson City in his van, he looked across the Klondike River and pointed out that no trace of Lousetown can be found today. A modern bridge crosses the Klondike several miles upriver where it joins Bonanza Creek. A paved two lane road winds up the narrow valley within sight of Bonanza Creek whose location has changed many times over the past century as giant dredges reprocessed many thousands of tons of soil along its course. Remarkably, one of the signs indicating distance from the original discovery is posted at 80 Below

*Pat standing at the site of #80 Below Discovery.*

Discovery, roughly coinciding with the Harris brothers' initial land purchase in the Klondike.

Visitors see a lot of abandoned mining equipment as they continue and our guide explained how it was all used. Where the Eldorado and Bonanza Creeks join, there is no trace of Grand Forks, the small city that sprouted there soon after the

*Pat with our guide David Taylor, examining an abandoned sluicebox.*

stampede began. But scattered along the edge of both creeks are dozens of modern prospectors, panning in hope of finding their personal fortune. One regular brings his laptop computer loaded with satellite photographic analysis of the creek beds. Another lines his sluicebox with Astroturf to facilitate entrapment of the smallest particles of gold. An amateur archeologist digs up garbage buried a century ago. Cans of cheap corned beef can be indicative of a poor site with little prior gold yield, but evidence of canned butter or tuna suggests that the claim holders were better off, perhaps because of a higher yielding claim. Many try to dig again where gold was found before.[21] And memories of the shooting of Dan McGrew are not that distant because a likely cause of violence today is still claim-jumping or sluicebox thievery.

*Panning for gold today at the union of Bonanza and Eldorado Creeks.*

And what do we know about the next rush for gold? Actually, it has already begun, albeit in a style of the late twentieth century. The Viceroy Resource Corporation[22] has invested $50 million Canadian in land and modern technology northeast of Dawson city. Their Brewery Creek operation uses the heap-leach method to chemically extract gold from the soil. This requires processing several tons of earth to yield a single ounce of pure gold.[23] Current production is 75,000 troy ounces per year ($22.5 million), equivalent to peak production in the year 1900.

Man's perpetual quest for gold endures!

*Pat tries her hand in Bonanza Creek where her grandfather staked his claim in 1897.*

# CHRONOLOGY

**October 6, 1862**- Robert b to Collingwood and Elizabeth Harris (3rd wife), Beaverton, Ontario.

**April 14, 1865** - Keturah b to Henry and Julia Whitney, Atherley.

**May 17, 1873** - Collingwood Harris contracts to build new narrows bridge.

**June 18, 1885** - Collingwood Harris dies, Atherley.

**November 24, 1886** - Julia Whitney and Thomas Harris marry, Atherley.

**July 3, 1888** - Keturah Whitney (KH) and Robtert Harris (RH) marry, Atherley.

**June 10, 1889** - Hazel b to RH and KH, Atherley.

**August 28, 1891** - Nina b to RH and KH, Atherley.

**March 22, 1893** - Donelda b to RH and KH, Atherley.

**August 17, 1896** - George Carmack's party discovers gold in Rabbit Creek, Yukon Territory.

**August 22, 1896** - New claimholders rename Rabbit Creek: Bonanza Creek.

**September 16 1896** - Nellie b to RH and KH, Vancouver.

**July 15, 1896** - Steamer *Excelsior* arrives in San Francisco.

**July 17, 1897** - Steamer *Portland* arrives Seattle - the rush is on!

**August 16, 1897** - Steamer *Islander* departs Victoria, 8:15 pm for Vancouver, Dyea.

**August 17, 1897** - Steamer *Islander* departs Vancouver, 8:20 am for Dyea.

**August 20, 1897** - Steamer *Islander* arrives Skagway - the hard work begins.

**October 22, 1897** - RH and Arthur Harris (AH) purchase one-half of #80 Below Discovery on Bonanza Creek for $3,000.

**November 2, 1897** - RH claims from crown #80b Below Discovery on Bonanza Creek.

**November 6, 1897** - AH claims #63 Above Discovery on Moosehide Creek.

**November 8, 1897** - AH claims #1 on Harris Creek at #61 on Orphir Creek.

**November .8, 1897** - RH claims #55 Above Discovery on Orphir Creek.

**November 13, 1897** - AH claims #30 Above Discovery on Deadwood Creek.

**November 18, 1897** - AH claims #23 on Left Fork Hunker Creek.

**November 30, 1897** - RH claims #57 Below Discovery on Henderson Creek.

**November 30, 1897** - AH claims #58 Below Discovery on Henderson Creek.

**December 13, 1897** - RH claims #32 Above Discovery on Enslay Creek.

**December 23, 1897** - RH sells #57 Below Discovery on Henderson for $275.

**December 28, 1897** - AH claims #1 on Tributary at #10 Above Discovery on Hunker Creek.

**January 5, 1898** - AH claims #31 on Tributary at #173 Above Discovery on Reindeer Creek.

**January 26, 1898** - RH purchases 1/2 of #76 Below Discovery on Hunker Creek for $700.

**February 17, 1898** - RH purchases 1/2 of #262 on Dominion Creek for $500.

**April 6, 1898** - RH returns to Vancouver on *Centennial* from Klondike.

**July 21, 1898** - KH and daughters visit family in Atherley.

**October 3, 1898** - RH sells #80b Below Discovery on Bonanza for $8,000.

**November 2, 1898** - AH & RH sell #80 Below Discovery on Bonanza for $8,000.

**February 28, 1899** - Mancel b to RH and KH, Atherley.

**May 11, 1899** - RH purchases seven lots including Capt. McDougal's hotel.

**January 26, 1901** - Oswald b to RH and KH, Atherley.

**April 14, 1901** - RH forfeits #262 on Dominion Creek to Nelson Soggs.

**May 30, 1901** - Grand opening celebration of *Peninsula House*.

**August 15, 1901** - Steamer *Islander* strikes iceberg in Lynn Channel and sinks.

**November 13, 1901** - RH and AH file suit against Pioneer Mining Company.

**January 6, 1902** - Pioneer quickly settles lawsuit in favor of RH and AH.

**January 21, 1904** - AH weds Kate Marshall, St Pauls Church, Vancouver.

**June 30, 1906** - Gretta b to RH & KH, Atherley.

**May 20, 1907** - Emma Harris McDougall dies, Atherley.

**February 21, 1908** - Stanley b to RH & KH, Atherley.

**April 20, 1911** - Capt. J.W. McDougall dies, Toronto.

**May 27, 1911** - Ida Elizabeth Harris Gaudaur dies, Atherley.

**February 5, 1914** - *Peninsula House* burns to the ground.

**October 28, 1915** - AH dies at age 55, Vancouver.

**December 11, 1922** - RH dies at age 60 after prolonged illness, Atherley.

**June 6, 1952** - KH dies at age 87, Atherley.

# ACKNOWLEDGEMENTS

Nothing short of interviewing Robert and Keturah themselves could have contributed more to the telling of their adventure. There were moments when we wished that we could have done just that. Next best were the many stories retold by their descendants. We are therefore grateful for the recollections passed down by daughters Hazel Harris, Nina Harris, Donelda Harris Barkey, Nellie Harris, Dorothy Harris Bunce, and Gretta Harris Lord; also sons Mancel, Oswald, and Stanley Harris.

When we began our research for this book in 1987, Pat and I interviewed Frances Kehoe Harris, wife of Oswald; Mary Hanguard and Helen Roe, nieces of Robert and Keturah; and the surviving grandchildren: Bernice Lord Birkett, Robert and Stanley Bunce, Donelda Barkey Fralick, Walter Lord Jr., Virginia Lord Luepschen, and Marian Barkey Polgar. Our sincere thanks go to all for their recollections, including three close versions of Robert's poem that we have blended into one that appears in Chapter 6.

Regrettably the *Peninsula House* fire in 1914 took most of the Yukon memorabilia with it. Documents still held by the family were therefore sparse. However we benefited enormously from articles in the *Orillia Packet*, found for us by Ken Macleod of Ottawa, Canada, a specialist in genealogy. He searched census records, land registries, even records of the Northwest Mounted Police. He directed us to the newspapers published in Victoria and Vancouver during the gold rush years. We are still amazed by his investigative skills and are most fortunate to have found him.

During our visits to Atherley, we were assisted by local historian Don Hunter and Orillia Librarian Frances Richardson who led us to an original of the *Peninsula House* brochure. Judy Gaudaur Savage, grandniece of Olympic sculler Jake Gaudaur, has provided valuable assistance throughout. She arranged an interview in Toronto with her aunt Stella Gaudaur Kehoe who, at age 95, provided memories of life in Atherley following Robert Harris's return from the Yukon. She gave us a vivid eyewitness account of the hotel fire.

At Seattle's Klondike Gold Rush Museum, we met Mark Blackburn and Todd Haskell who provided valuable information and access to National Park Service's excellent video collection. In Victoria, B.C., the Maritime Museum staff directed us to a replica of the steamer *Islander*. Nellie Harris's birth record was retrieved from the Division of Vital statistics, Ministry of Health.

The City of Vancouver offered many resources essential for completion of our research. In the Maritime Museum, more records of the *Islander*, its history and tragic end, were found for us by Joan Thornley. At the Public Library, we were aided by Maureen Matthews in the Northwest History Room where an important photographic archive exists. The Vancouver Archives maintains a microfilm newspaper collection of the gold rush years, and located deed of purchase documents for the Keefer Street property.

I am indebted to my surgical colleague, Dale Birdsell, who took time during a Calgary visit to drive up the Bow River Valley, alongside the CPR's original westward line to Kicking Horse Pass. There, we stood by the abandoned tracks for the steep descent towards British Columbia. Later, we watched modern freight trains descend gently through the spiral tunnels that now exist at Kicking Horse Pass.

Doug Charles at the Tongass Historical Museum in Ketchikan, Alaska taught us all about gasboats and the diverse fishing fleet used in Pacific Northwest waters at the turn of the century. Sally Robinson at the superb State Historical Museum in Juneau provided us with access to their extensive collection of Tlingit, Athabascan, and gold rush artifacts.

In Skagway, National Park Service historian Karl Gurcke was generous with his time, as was Park ranger Brad Belo, whose Parks Canada counterpart Odette Lloyd, provided details of topographic conditions and artifacts along the Chilkoot Pass trail, then and now. Emily Klimek at the Skagway Visitors Center was able to provide cruise ship and tourist visitor statistics for modern Skagway.

Very sincere thanks go to the many specialists who assisted us at the Yukon Territorial Archives in Whitehorse, among them Dolores Smith, Diane Chishold, and Fay Tangerman. Not only did they inspire our project by finding Robert A. Harris' name in the Bonanza Creek book, they also taught us how to search the other creek books for the evidence we sought. We admire them to this day for their skill, and appreciate their patience with us. Another fine resource in Whitehorse is the MacBride Museum where Brenda Carson gave assistance.

In Dawson City, we received invaluable help from many fine people and we are grateful to them all. Michael Gates, author of *Gold at Fortymile*, is now Curator of Collections, Klondike National Historic Sites for Parks Canada. Thank you Michael for your book, and for offering suggestions after reviewing selected chapters of *Lead Pencil Miner*. John Gould at the Dawson City Museum directed us to specific mining records in their carefully acquired archives. We were also guided by Candy Evans, Shabira Tamachi, and Mac Swackhammer who now directs the Museum.

We thank Dick North at the Jack London Center in Dawson City for his fascinating stories and encouragement. He later referred us to Winnie at the Jack London Book Store in Glen Ellen, California who provided us with a copy of the Fred Thompson diary that documents London's Yukon journey.

David "Buffalo" Taylor of Gold City Tours will always know how thrilled we were with his guidance through the gold fields near Dawson City.

Many photographs were made available to us by the Special Collections Library of the University of Washington. Marilyn Blaisdell kindly provided the photograph of abandoned ships in San Francisco harbor.

Special thanks go to Burton Brockett, Rikki Ellwood, and Eldonna Lay. Burt's graphics, book design and countless other production details were critical to completion of the project. I have learned from him exactly how a book is made. Rikki assisted with manuscript preparation and proofreading. Eldonna's editorial guidance was essential for developing the story narrative.

Finally, I want to thank my wife, Pat, seventh grandchild of Robert and Keturah Harris. What was to have been a pamphlet has become a book, largely because of her persistence and thoroughness. What fun it has been!

# CHAPTER NOTES

## Chapter One - *Man's Endless Search for Gold*

**1**. First use of coins by Lydians cited in the writings of Herodotus who was apparently not aware of a Chinese precedent. Herodotus, Book I, William Beloe translation. McCarty & Davis. Philadelphia, 1844, p31.

**2**. Rulers of every nation that followed became enamored of seeing their facial likeness on coins with inherent value. After the death of Caligula, his money was called in and melted down so his features could be forgotten. J.K. Galbraith. *Money, Whence it Came, Where it Went*, p7-17.

**3**. Croesus - 561-546 BC. His monetary innovations are discussed by R.E. Doty in *Money of the World*.

**4**. Matthew 26:14-16: Judas Iscariot received thirty silver coins for identifying Jesus among the twelve disciples.

**5**. Despite these safeguards, rulers could devalue a coin anytime they wished by reducing the quantity of silver and replacing it with a metal of lesser value. After the financial pressures of the Punic Wars, the silver content of basic coins was reduced to two percent. Thus money could be reliable yet scarce. Or it could be abundant and worth little. J.K. Galbraith. *Money, Whence it Came, Where it Went, p7-17.*

**6**. Visitors to Prague's Hradcany Castle can still view the former workshops of these alchemists.

**7**. The Atlantic Ocean was referred to as the 'Ocean Sea' at that time.

**8**. Columbus required nearly 100 crewmen for his three ships: Nina, Pinta, and Santa Maria. The Prisons offered him many and he enticed the rest with visions of gold as described by Marco Polo. S. E. Morison, *Admiral of the Ocean Sea.*

**9**. The destination Columbus sought was Japan, which he calculated to be 3,000 miles east of Lisbon. Flying today by commercial air from Lisbon to Tokyo is a journey exceeding 10,000 miles.

**10**. The only major deposits of gold found in the western hemisphere within the latitudes sailed by Columbus were in Costa Rica many years later. G. Granzotto. *Christopher Columbus.*

**11**. During the Aztec conquest by Hernan Cortes (1519-1521), Montezuma was most willing to oblige his Spanish invaders by directing them to sources of gold. The Aztecs did not yet value gold as their conquerors did; jade was a more precious commodity in their eyes. H. Thomas. *Conquest: Montezuma, Cortes, and the Fall of Old Mexico.*

**12**. Known by historians of financial folly as the Banque Royale scandal, speculators in Paris formed 'The Mississippi Company'. Thousands of shares were sold to pursue gold deposits falsely presumed to exist in North America. The proceeds went to the government for its debts instead of in search for the undiscovered gold. J.K.Galbraith. *A Short History of Financial Euphoria*, p37-42. Author's note: As Chapter One was being written, Bre-X Minerals Ltd. claimed gold discoveries in Indonesia that proved false because of fabricated ore analysis. While this chapter was being proofread, a Connecticut-based hedge fund, Long Term Capital Investment, received a bailout from major banks because they tried to leverage a $2 billion investment into a $100 billion play - and failed!

**13**. In the eighteenth century, these manifestations of business cycling were called 'crises', in the nineteenth century, 'panics', and in the early twentieth century, 'depressions'. More recently they are called 'recessions' or 'economic corrections'.

**14**. This discovery came just nine days after Mexico signed the Treaty of Guadalupe Hidalgo, thereby relinquishing the California Territory and all of its gold to the United States. J.S. Holiday. *The World Rushed In.*

**15**. Inexplicably, the phrase 'Pike's Peak Gold' was coined even though the South Platte discovery was 80 miles to the north. Hundreds of men set off with 'Pike's Peak or Bust' signs emblazoned on their wagons. Little gold was ever found there.

**16**. In 1893, there were only 22 industrial issues traded on the New York Stock Exchange, one of them a new company called General Electric.

**17**. Grover Cleveland had been elected Mayor of Buffalo in 1881, and Governor of New York State in 1882.

**18**. General Electric survived the Panic of 1893, and is the only industrial corporation of that era still in existence today.

## Chapter Two - *Looking Westward*

**1**. The last spike in the United States transcontinental railroad was driven on May 10, 1869 at Promontory Point, Utah. The following message was telegraphed to both coasts simultaneously: "Done!" In Philadelphia the Liberty Bell rang, and in San Francisco a banner read, "California Annexes the United States!".

**2**. Known geologically as the 'precambrian shield, or Canadian shield, its penetration by the railroad required three dynamite factories producing $7.5 million of the explosive, making the central third of the entire CPR project more costly per mile than crossing the Rocky Mountains. P. Benton. *The Great Railway*, p329.

**3**. Ibid., p19

**4**. Collingwood Harris born 1809 in Port Hope, Ontario; married Elizabeth Murphy Oct. 13, 1885 in Toronto; died June 13,1885 in Atherley at the age of 76.

**5**. Thomas Harris born Mar. 12, 1858, died July 11, 1923; Arthur Harris born 1860, died 1915; Robert Harris born Oct. 6, 1862, died Dec 11, 1922; Shuter Harris born 1870, died 1927.

**6**. Emma Harris born Aug. 15, 1856, died May 20, 1907; Ida

Harris born Apr. 22, 1867, died May 27, 1911; dates of birth and death for Rachel and Martha not known.

**7**. Julia Whitney and Thomas Harris married Nov. 12, 1886.

**8**. Keturah Whitney and Robert Harris married July 3, 1888.

**9**. 'Gasboats' were powered by 5 hp inboard engines that could run on various oil derived fluids like kerosene, benzene, or napthalene. Larger draft vessels of that day required steam engines fueled by coal. From documents provided by Douglas Charles at Tongass Historical Museum in Ketchikan, Alaska.

**10**. Mortgage #7457A recorded January 6, 1890 in the Superior Court of British Columbia for purchase of Lot 13, Block 86, Part of District Lot 196.

**11**. English Bay Canning Company incorporated May 5, 1898 for $50,000 (500 shares at $100 each). At the peak of operation, it employed 300 hands, 300 fishermen. It faced English Bay at the foot of Trutch Street. In 1934, its exact site was still identifiable by the enormous pile of tin cuttings left behind from the fabrication of salmon cans.

**12**. Ralston, K. *Patterns of Trade and Investment on the Pacific Coast* 1867-1892, p 37-54.

**Chapter Three** - ***"A Ton of Gold!"***

**1**. Gates, Michael, *Gold at Fortymile Creek: Early Days in the Yukon.*

2. Ibid., p5.

**3**. Ibid., p8.

**4**. Ibid., pp 68-78.

**5**. Ibid., pp 41-43.
**6**. Ibid., pp 68-71.

7. Ibid., pp 115-119.

**8**. Local Indian pronunciation of 'tl' and 'thr' is like 'kl'. Thus, we should say 'Klingit', not 'Tlingit' as it looks, which explains why the Thronduik River became known throughout the world as 'Klondike'.

**9**. Gates, Michael, *Gold at Fortymile Creek: Early Days in the Yukon.*

**10**. Berton. Pierre, *The Klondike Fever*, pp 51-61.

**11**. Matthews, Richard, *The Yukon: Rivers of America Series.*

**12**. Berton, Pierre, *The Klondike Fever*, pp 90-95.

**13**. Adney, Tappan *The Klondike Stampede*, pp193-198.

**Chapter Four** - ***Heading for the Yukon***

**1**. Discerning readers will recall that certain ancient forms of communications like smoke signals also traveled at the speed of light but only for limited distances.

**2**. Nordstrom, John, *The Immigrant* in 1887.

**3**.Ibid, p 27

**4**. The *Islander* was built in Glasgow, Scotland in 1888. It sailed to British Columbia via the Straits of Magellan.

**5**. Adney, Tappan *The Klondike Stampede.*

**6**. Prior to the *Portland's* arrival, Seattle's population was 40,000 and the city did $300,000 in trade each year. One year later, its commercial trade grew to $25 million a year. In 1910 its population reached 210,000.

**7**. Berton, Pierre. *The Klondike Fever*, pp 120-122.

**8**. Adney, Tappan. *The Klondike Stampede*, pp 11-15.

**9**. Ibid., p 22.

**10**. The *Islander* was owned by the Irving family until 1901 when it was sold to the CPR.

**11**. Today Juneau is the colorful and picturesque capital of Alaska.

### Chapter Five - *The Race for Dawson City*

**1**. Adney, Tappan. *The Klondike Stampede*, p 71-73.

**2**. Matthews, Richard. *The Yukon: Rivers of America Series.*

**3**. Pronounced 'Klinkit".

**4**. The argonauts faced depletion of their monetary resources if they committed to porters for all their goods. Yet they faced the harshness of winter and exhaustion if they carried their own goods over the pass.

**5**. This is the scene most frequently recorded and displayed for historical recollection of the Yukon gold rush.

**6**. McIllree, J.H. Annual Report of the Asst. Commissioner, Northwest Mounted Police, 1897.

**7**. Nordstrom, John W. *The Immigrant* in 1887.

**8**. Adney, Tappan. *The Klondike Stampede.*

### Chapter Six - *Lead Pencil Miner*

**1**. Grant Appl. #860 to The Crown filed October 23, 1897 for purchase of one-half interest in #80 Below Discovery on Bonanza Creek from H. McCullough by Arthur J. and Robert A. Harris for $3,000 ($1,400 downpayment., balance due July 1, 1898).

**2**. Ogilvie, William. *Early Days in the Yukon.*

**3**. Grant Appl. #2094 to The Crown filed November 5, 1897 for #80b Below Discovery on Bonanza Creek by Robert A. Harris.

**4**. Grant Appl. #2317 to The Crown filed November 10, 1897 for #63 Above Discovery on Moosehide Creek by Arthur J. Harris.

**5**. Grant Appl. #2294 to The Crown filed November 19, 1897 for #1 on Harris Creek, tributary at #61 Above Discovery on Orphir Creek by Arthur J. Harris.

**6**. Grant Appl. #2268 to The Crown filed November 9, 1897 for #55 Above Discovery on Orphir Creek by Robert A. Harris.

**7**. Grant Appl. #2739 to The Crown filed November 20, 1897 for #30 Above Discovery on Right Fork of Deadwood Creek by Arthur J. Harris.

**8**. Grant Appl. #2740 to The Crown filed December 7, 1897 for #23 on Left Fork of Hunker Creek by Arthur J. Harris.

**9**. Grant Appl. #4106 to The Crown filed December 8, 1897 for #1 on tributary at #10 Above Discovery on Hunker Creek by Arthur J. Harris.

**10**. Grant Appl. #3230 to The Crown filed December 8, 1897 for #58 Below Discovery on Henderson Creek by Arthur J. Harris.

**11**. Grant Appl. #3204 to The Crown filed December 7, 1897 for #57 Below Discovery on Henderson Creek by Robert A. Harris.

**12**. Grant Appl. #3418 to The Crown (date of filing unknown) for #32 Above Discovery on Enslay Creek by Robert A. Harris.

**13**. Grant Appl. #1795 to The Crown filed February15, 1898 for sale of #57 Below Discovery on Henderson Creek to F.A. Keene and B.A. Burton on Dec. 28, 1898 by Robert A. Harris.

**14**. Grant Appl. #4407 to The Crown (date of filing unkown) for #31 on tributary at #173 Above Discovery on Reindeer Creek by Arthur J. Harris.

**15**. Grant Appl. #1679 to The Crown filed January 31, 1898 for purchase of one-half interest in #76 Below Discovery on Hunker Creek from C.L. Berg by Robert A. Harris.

**16**. Grant Appl. #1990 to The Crown filed February 28, 1898 for purchase of one-half interest in #262 Below Discovery on

Dominion Creek from Mr. P. Mahon by Robert A. Harris.

**17**. This version is a blending of recollections from several family members.

### Chapter Seven - *Keturah's Story*

**1**. Hazel Harris born June 10,1889, Atherley, Ontario.

**2**. Nina Harris born August 28, 1891, Atherley, Ontario.

**3**. Donelda Harris born March 22, 1893, Atherley, Ontario.

**4**. Nellie Harris born September 16, 1896, Vancouver, British Columbia.

**5**. Chandonnet, A. *"Ma" Pullen's Kitchen.*

**6**. Zanjani, S. *A Mine of Her Own: Women Prospectors in the American West.*

**7**. The story, as related to a reporter in Orillia is complicated, but it is most likely that Robert had buried surplus provisions en route, perhaps after he purchased the marooned boat together with its supplies. Flour was scarce in Dawson City whereas game was plentiful. Robert very likely financed his journey home by agreeing to carry bear meat out to ready buyers on the coast while his Indian guide took flour from the buried cache back to Dawson City.

**8**. "A Vancouverite Returns - A Very Rosy Prospect for All," Vancouver Daily News-Advertiser. April 7, 1898.

### Chapter Eight - *Return to the Yukon*

**1**. "A Vancouverite Returns - A Very Rosy Prospect for All." *Vancouver Daily News-Advertise*r, April 7, 1898.

**2**. Notice in *Orillia Packet*, July 21, 1898.

**3**. Satterfield, A. *Chilkoot Pass*. Ch. 16: Railroad Builders. p154-157.

**4**. Berton, Pierre. *The Klondike Fever.* p161-165, 333-365.

**5**. Adney, Tappan. *The Klondike Stampede.*

**6**. Grant Appl. #6007 filed October 3, 1898 for sale of #80b Below Discovery on Bonanza Creek to A.J. Mangold, agent for Pioneer Trading Company, London by Robert A. Harris for $8,000.

**7**. Syndicates backed by enormous capital investment brought several dredges to the gold bearing creeks and reworked many thousands of tons of soil.

**8**. Grant Appl. #7323 filed November 2, 1898 for sale of quarter interest in #80 Below Discovery on Bonanza Creek to William Bradley by Robert and Arthur Harris for $8,000.

### Chapter Nine - *Back to Atherley*

**1**. Collingwood Harris obituary. *Orillia Packe*t, June 26, 1885.

**2**. Listing of local hotels and resorts in The Northern Light, May 13, 1870.

**3**. McDougall obituary in *Orillia Packet*, April 20, 1911.

**4**. Announcement in *The Orillia Daily Times*, July 26, 1889.

**5**. Articles in the *Orillia Packet*, January 30, 1912.

**6**. Announcment of *Peninsula House* grand opening appeared in Orillia Packet, May 30, 1901.

### Chapter Ten - ***Aftermath***

**1**. Gates, Michael. *Gold at Fortymile: Early Days in the Yukon.*

**2**. Ibid., p148.

**3**. Ibid., p148.

**4**. Ibid., p149.

**5**. Ibid., p148.

**6**. Benet, W. R. *The Reader's Encyclopedia.* Thomas Crowell. New York, 1965.

**7**. Nordstrom, John W. *The Immigrant in 1887.*

**8**. Adney, Tappan. *The Klondike Stampede.*

**9**. Ibid., p145.

**10**. Vancouver's Maritime Museum has a scale model of the *Islander* and a file with accounts of its sinking as well as later salvage efforts.

**11**. Total production during the California Gold Rush was 820 tons, or 24 million ounces. It is estimated that the earth's crust contains ten billion tons of gold, only a small percentage of it being placer gold. So far, history records a cumulative harvest of 88,000 tons, an amount that would fill a cube with 18-yard sides.

**12**. Arthur filed his forwarding address with the Dawson City Post Office, a record that exists to this day in the Dawson City Museum.

**13**. All liens on Harris property recorded by Registry Office, County of Simcoe, Ontario, Canada.

**14**. McDougal obituary, *Orillia Packet*, April 20, 1911.

**15**. Donelda Harris married Walter Barkey in Toronto, Canada.

**16**. Dorothy Harris married Clarence Bunce in Atherley, Ontario.

**17**. Gretta Harris married Walter Lord Jr. in Buffalo, New York.

**18**. Oswald Harris married Frances Kehoe in Atherley, Ontario.

**19**. The Harris estate was probated in 1987 and the proceeds divided among eight surviving grandchildren: Bernice Birkett, Robert Bunce, Stanley Bunce, Patricia Fisher, Donelda Fralick, Walter Lord Jr., Virginia Luepschen, and Marian Polgar.

**Epilogue - *In the Yukon Today***

**1**. Klondike Gold Rush National Historical Park Museum, 117 S. Main Street at First Avenue, Seattle, Washington (in a building that housed the Union Trust Bank in 1897). (206) 553-7220.

**2**. Seattle Post-Intelligencer, Wall Street and Sixth Avenue. (206) 448-8066.

**3**. University of Washington Library, Special Collections, Seattle, Washington. (206) 543-1929.

**4**. Empress Hotel, 721 Government Street, Victoria, B.C. V8W 1W5. (250) 384-8111.

**5**. Maritime Museum of British Columbia, 28 Bastion Square, Victoria, B.C. V8W 1H9. (250) 385-4222.

**6**. Named for a talkative saloon keeper, 'Gassy' Jack Deighton.

**7**. In addition to the Hudson Bay Company Building, these include the Byrne's block at 2-8 Water St., Terminal Hotel at 28-32 Water St., and Greenshield's Building at 335-347 Water St.

**8**. Vancouver Public Library.(604) 331-3600

**9**. British Columbia Archives.(604) 736-8561

**10**. Vancouver Maritime Museum, 1905 Ogden Street: (604) 257-8300.

**11**. Alaska Marine Highway: (800) 642-0066.

**12**. Golden North Hotel, Broadway and Third Street, Skagway, Alaska.(907) 983-2451

**13**. Contact the National Park Service, U.S. Dept. of Interior, Klondike Gold Rush National Historical Park, P.O. Box 517, Skagway, Alaska, for more detailed information about the Chilkoot Pass trail.

**14**.MacBride Museum, First Avenue and Wood Street, Whitehorse, Y.T. (867) 667-2709.

**15**. Yukon Archives, Whitehorse, Y.T. Canada, Y1A-2C6 (867) 667-2709.

**16**. Air North, (800) 764-0407 from U.S.; (800) 661-0401 from Canada. Until recently, restored DC-3's were used between Whitehorse and Dawson City, a comfortable and exciting 1 hour 40 minute ride. They have since been replaced by more modern craft.

**17**. Berton, Pierre. *Drifting Home.* McLelland & Stewart. Toronto, 1973.

**18**. Robert Service was born in Preston, England. He first traveled to British Columbia in 1896 but did not come to the Yukon until 1904. Based on his observations of life in Dawson City, he wrote his first poem there: *The Shooting of Dan McGrew.* The rest is history!

**19**. Dawson Museum and Historical Society, P.O. Box 303, Dawson City, Y.T., Canada Y0B-1G0. (867) 993-5291.

**20**. Gold City Tours, Dawson City. Proprietor: David 'Buffalo' Taylor: (867) 993-5175.

**21**. "Yukon Miners Mix New Methods With Old in Prospecting for Gold", Wall Street Journal. April 17, 1987.

**22**. Viceroy Resources Corporation, Vancouver, B.C. (604) 688-9780.

**23**. "Mining's Massive Scale", San Diego Union-Tribune.

# REFERENCES

## Primary Sources

Adney, Tappan. *The Klondike Stampede*. Harper Bros. New York, 1900.

Berton, Pierre. *Drifting Home*. Mclelland & Stewart Ltd. Toronto, 1973.

Berton, Laura B. *I Married the Klondike*. Mclelland & Stewart. Toronto, 1954.

Nordstrom, John W. *The Immigrant in 1887*. Dogwood Press. 1950.

Ogilvie, William. *Early Days in the Yukon*. Ottawa, 1913.

Steele, Samuel B. *Forty Years in the Yukon*. London, 1915.

Thompson, Fred. *Diary of Fred Thompson*. Jack London Bookstore. Glen Ellen, Calif .

## General References

Atkin, John. *Strathcona: Vancouver's First Neighborhood*. Whitecap Books. Vancouver, 1994.

Berton, Pierre. *The Great Railway*. McClelland & Stewart. Toronto, 1974.

Berton, Pierre. *The Klondike Fever*. Carroll & Graf. New York, 1958.

Bolotin, Norman. *Klondike Lost: A Decade of Photographs*. Alaska Northwest Publishers. Anchorage, Alaska, 1980.

Bolotin, Norman. *A Klondike Scrapbook*. Chronicle Books. San Francisco, 1987.

Boorstin, Daniel J. *The Discoverers*. Random House. New York, 1983.

Braasch, Barbera. *California's Gold Rush Country*. Johnson Int.. Medina, Wash., 1996.

Brittanica, The New Encyclopedia. Encyclopedia Brittanica Inc. Chicago, 1985.

Clifford, Howard. *The Skagway Story*. Alaska Northwest Publishers. Anchorage, Alaska, 1975.

Cohen, Stan. *The White Pass and Yukon Route*. Pictorial Histories Publishing. Missoula, Montana, 1980.

Doty, Richard. *Money of the World*. Grosset & Dunlop. New York, 1978.

Galbraith, John Kenneth. *The Great Crash:1929*. Houghton, Mifflin Co. Boston, 1954.

Galbraith, John Kenneth. *Money, Whence it Came, Where it Went.* Andre Deutsch. London, 1975.

Galbraith, John Kenneth. *A Short History of Financial Euphoria.* Viking. New York, 1990.

Gates, Michael. *Gold at Fortymile: Early Days in the Yukon.* Univ. British Columbia Press. Vancouver, 1994.

Granzotto, Gianni. *Christopher Columbus.* Univ. Oklahoma Press. Norman, Oklahoma, 1985.

Holliday, J.S. *The World Rushed Inn.* Simon & Schuster. New York, 1981.

Lopez, Francisco. *Cortes, Life of the Conqueror.* Univ. California Press. Berkeley, 1964.

Matthews, Richard. *The Yukon: Rivers of America Series.* Holt Rinehart. New York, 1968.

Mayer, Melanie. *Klondike Women.* Ohio Univ, Press. Columbus, 1989.

McElvane, R.S. *The Great Depression: America 1929-1941.* Times Books. New York, 1984.

Morison, Samuel Eliot. *Admiral of the Ocean Sea.* Northeastern Univ. Press. Boston, 1942.

Morison, Samuel Eliot. *Christopher Columbus: Mariner.* New American. New York, 1942.

Rorbaugh, M.J. *Days of Gold.* Univ. California Press. Berkely, 1997.

Rosenberg. N. and Birdzell, L.E. *How the West Grew Rich: Economic Transformation of the Industrial World.* Basic Books. New York, 1986.

Satterfield, Archie. *Chilkoot Pass.* Alaska Northwest Publishers. Anchorage, Alaska. 1973.

Service, Robert. *The Complete Poems of Robert Service.* Blakiston. Philadelphia, 1944.

Thomas, Hugh. *Conquest: Montezuma, Cortes, and the Fall of Mexico.* Simon & Schuster. New York, 1993.

Wilcox, John. *Vancouver.* Houghton Mifflin Co. New York, 1993.

Walker, Franklin. *Jack London and the Klondike.* The Huntington Library Press. San Marino, California, 1966.

Zanjani, Sally. *A Mine of her Own: Women Prospectors in the American West, 1850-1950.* Univ. Nebraska Press. Lincoln, Nebraska, 1997.

**Articles, Pamphlets**

Belcher, H.A. All About the Klondike with Full Particulars by a Returned Miner.

Chandonnet, Ann. "Ma" Pullen's Kitchen. Alaska. Sept. 1997, p80.

Guide to Ontario Land Registry Records. The Ontario Genealogical Society. Toronto, 1994.

Index to Creeks and Tributaries, Series 10 Mining Recrders Records, Record Books for Placer Mining Claims: 1896-1969. Yukon Archives. Whitehorse, Y.T. 1989.

Man and His Gold. Prepared by The Gold Information Center.

McIlree, J.H. Annual Report of the Asst. Commissioner, Northwest Mounted Police, 1897.

Neufeld, David. Chilkoot Trail. 1993.

Ralston, K. Patterns of Trade and Investment on the Pacific Coast, 1867-1892. British Columbia Studies Vol 1, p37-54.

Sack, Doug. Gold: A Brief History of Dawson City and the Klondike. The Monte Carlo Ltd., 1979.

Thirkell, Fred. Vancouver's Past. Gordon Soules Book Publishers Ltd. Vancouver.

White, Peter T. The Eternal Treasure: Gold. National Geographic. Jan., 1974, pp 1-51.

**Newspaper Archives Consulted**

Orillia Packet and Times.

The Orillia Daily Times

Seattle Post-Intelligencer.

Vancouver Daily News-Advertiser.

Vancouver Semi-Weekly World.

**Video Recordings**

Klondike: The American Experience, PBS.

Scams, Schemes, Scoundrels, Discovery Channel.

White Pass and Yukon Route, WPYR, Skagway, Alaska.

**Photographic Archival Sources**

Orillia Library.

University of Washington.

Vancouver Archives.
Vancouver Maritime Museum.
Vancouver Public Library.

**Photo Credits**

| | |
|---|---|
| Cover | *SS Islander*, British Maritime Museum |
| vii | Klondike headline, *The Seattle Post-Intelligencer.* |
| 6 | Boats, San Francisco Harbor. Marilyn Blaisdell Collection. |
| 7 | English Bay Cannery, Victoria Archives |
| 21 | G. Carmack, Skookum Jim, Tagish Charley, Robert Henderson. University of Washington-Hegg, 1894. |
| 26 | *SS Excelsior*, San Francisco Harbor. University of Washington. |
| 30 | Cooper & Levy, Seattle. University of Washington-Curtis 1897 |
| 35 | Steamer at Vancouver Wharf, Maritime Museum, Vancouver, B.C. - #2544. |
| 36 | *SS Islander*, Provincial Archives of B.C. - #217-A79. |
| 41 | Supplies at Skagway. Yukon Archives-Vogee 1898. |
| 42 | Barges at Dyea. Yukon Archives- Vogee 1898. |
| 45 | The scales at Chilkoot Pass. University of Washington-Hegg 1896. |
| 45 | Chilkoot Pass, view down. University of Washington-Hegg 1896. |
| 46 | Supplies atop Chilkoot. University of Washington-Hegg 1896. |
| 47 | Supplies at Long Lake. Yukon Archives - #136 1898 |
| 48 | Tent city, Lake Bennett. University of Washington. |
| 50 | Piloting through Miles Canyon. Yukon Archives - #3906. |
| 52 | Front Street, Dawson City. University of Washington-Hegg 1896. |
| 57 | Claimsites on Bonanza Creek. University of Washington. |
| 58 | Miners digging in shaft. University of Washington-Hegg 1897. |
| 58 | Gold Assayer's Office, Dawson City. University of Washington- L&D 1899. |
| 70 | Broadway, Skagway. Dedman Photos. |
| 75 | White Pass trestle. University of Washington-Hegg 1899. |

# INDEX